Springer Theses

Recognizing Outstanding Ph.D. Research

For further volumes:
http://www.springer.com/series/8790

Aims and Scope

The series "Springer Theses" brings together a selection of the very best Ph.D. theses from around the world and across the physical sciences. Nominated and endorsed by two recognized specialists, each published volume has been selected for its scientific excellence and the high impact of its contents for the pertinent field of research. For greater accessibility to non-specialists, the published versions include an extended introduction, as well as a foreword by the student's supervisor explaining the special relevance of the work for the field. As a whole, the series will provide a valuable resource both for newcomers to the research fields described, and for other scientists seeking detailed background information on special questions. Finally, it provides an accredited documentation of the valuable contributions made by today's younger generation of scientists.

Theses are accepted into the series by invited nomination only and must fulfill all of the following criteria

- They must be written in good English.
- The topic should fall within the confines of Chemistry, Physics, Earth Sciences, Engineering and related interdisciplinary fields such as Materials, Nanoscience, Chemical Engineering, Complex Systems and Biophysics.
- The work reported in the thesis must represent a significant scientific advance.
- If the thesis includes previously published material, permission to reproduce this must be gained from the respective copyright holder.
- They must have been examined and passed during the 12 months prior to nomination.
- Each thesis should include a foreword by the supervisor outlining the significance of its content.
- The theses should have a clearly defined structure including an introduction accessible to scientists not expert in that particular field.

Xiaofei Yu

Material Cycling of Wetland Soils Driven by Freeze–Thaw Effects

Doctoral Thesis accepted by
University of Chinese Academy of Sciences (former
Graduate University of Chinese Academy of Sciences),
China

 Springer

Author
Dr. Xiaofei Yu
Northeast Institute of Geography
  and Agroecology
Chinese Academy of Sciences
Changchun
People's Republic of China

Supervisor
Prof. Dr. Guoping Wang
Northeast Institute of Geography
  and Agroecology
Chinese Academy of Sciences
Changchun
People's Republic of China

ISSN 2190-5053   ISSN 2190-5061   (electronic)
ISBN 978-3-642-42918-7   ISBN 978-3-642-34465-7 (eBook)
DOI 10.1007/978-3-642-34465-7
Springer Heidelberg New York Dordrecht London

**Parts of this thesis have been published
in the following articles**

Xiaofei Yu, Yuanchun Zou, Ming Jiang, Xianguo Lu, Guoping Wang. Response of soil constituents to freeze-thaw cycles in wetland soil solution. Soil Biology & Biochemistry, 2011, 43(6): 1308-1320. (Reproduced with Permission)

Xiaofei Yu, Yuxia Zhang, Yuanchun Zou, Hongmei Zhao, Guoping Wang. Sorption and desorption of ammonium in wetland soils subject to freeze-thaw. Pedosphere, 2011, 21(2): 251-258. (Reproduced with Permission)

Xiaofei Yu, Yuxia Zhang, Hongmei Zhao, Guoping Wang. Freeze-thaw effects on sorption/desorption of dissolved organic carbon in wetland soils. Chinese Geographical Science, 2010, 20(3): 209-217. (Reproduced with Permission)

Xiaofei Yu, Guoping Wang. Field simulation the effects of global warming on the freeze-thaw process of the freshwater marsh in the Sanjiang Plain, Northeast China. Environmental Pollution and Public Health (EPPH2008). Shanghai, 2008. (Reproduced with Permission)

Supervisor's Foreword

Ongoing global warming may cause permafrost thaw to accelerate in high-latitude regions. Meanwhile, thawing permafrost will likely accelerate global warming in the coming decades, driven by increasing greenhouse gases production and emission from soil partially controlled by freeze–thaw cycles. Although freeze–thaw cycles can alter soil physical properties and microbial activity, their overall impact on soil functioning remains unclear. The question of how freeze–thaw events affect soil processes has attracted more and more attention in recent years, especially as climate change scenarios predict that the frequency of such events might increase in many ecosystems in the future.

Wetlands are among the most important ecosystems on Earth as they store more carbon than other terrestrial ecosystems, despite the fact that they cover only approximately 7 % of the total land surface. Peatland is an important type of wetlands and estimates of carbon stored in the global boreal peatlands range from 20 to 35 % of global terrestrial carbon. Many of the world's northern peatlands are underlain by rapidly thawing permafrost. The permafrost prevents soil drainage and thus deprives the oxygen in soil environment, which plays a crucial role in the accumulation and decay of organic matter. Freezing and thawing of soils occur regularly in high latitudes above 45°N as a result of the climatic conditions. Freezing and thawing of soils may affect the turnover of soil organic matter and thus the losses of carbon and nitrogen from soils. Soil freeze–thaw cycles can substantially affect soil carbon and nitrogen cycling, and deserve special consideration for wetlands in the winter-cold zone.

This thesis was written by Dr. Xiaofei Yu, and addresses the freeze–thaw of wetland soils with regard to several aspects: the dynamics of dissolved carbon and nitrogen in wetland soils, overlying ice and water during the freeze–thaw period; the freeze–thaw effects on the sorption/desorption of dissolved organic carbon and NH_4^+, and the mineralization of organic carbon and organic nitrogen; the dynamics of accumulation and release of dissolved carbon and nitrogen in intact wetland soils. Therefore, the freeze–thaw effects on the accumulation and release process of carbon and nitrogen in wetland soils were systematically investigated in the dissertation. It is a good step toward the investigation of wetland biogeochemical process in permafrost and seasonal freeze– thaw areas. It is also developing strategies aimed at global warming effects on the accumulation and release of carbon and nitrogen in wetlands. The studied process might be important in many wetland soils of cold regions. Globally, investigations in terms to cold climate processes are relatively sparse and need more attention.

This thesis also compared the effects of successive events of freezing and thawing the release of nutrients in natural and arable soils in northern China. The documentation of this process is an interesting and important contribution to the understanding of nutrient and release in northern wetlands. The results provide information on the timing of nutrient release related to freezing and thawing in natural versus arable soils, and has implications for the timing of nutrient application in farm fields in relation to water quality protection.

The major value of this work is twofold: it constitutes a useful methodological baseline for accumulation and release of carbon and nitrogen in wetland soils under freeze–thaw conditions, and it presents a novel approach in soil freeze–thaw cycle experiment, which is applicable not only for wetland soils but also for other soils. The thesis has already spawned several journal papers, and opens new perspectives for scientific problems in freeze–thaw processes and soil chemistry, biogeochemistry of wetlands, and global warming.

Changchun, April 2012   Prof. Dr. Guoping Wang

Contents

Chapter 1
Introduction

1.1 Background and Significance of the Research

1.1.1 Research Background

Climate is the main determinant of ecosystem conditions in a region. For many years, research has focused on the effects of climate factors such as water and heat on ecosystems, whereas research on the effects of freeze–thaw on ecosystems is relatively rare. Freeze–thaw is an abiotic stress applied to soils and is a natural process in medium and high latitudes. Freeze–thaw can change soil physical and chemical properties, and therefore affect the circulation and loss of nutrients in soils. In recent years, although some studies have investigated freeze–thaw effects on soil physical and chemical properties, most have focused on agricultural soils or forest soils, while little attention has been paid to the effects of freeze–thaw on carbon and nitrogen accumulation and release in wetland soils.

Freeze–thaw can change the size and stability of soil aggregates. When soils are frozen, disruptive forces affecting aggregates come from the expansion of ice crystals in soil pores. Expansion of these ice crystals breaks particle-to-particle bonds and splits the aggregates into smaller ones. Large particles are broken and the resulting fine particles have a tendency to re-form into medium size particles (Bullock 1988). For aggregates themselves, freeze–thaw applies a more disruptive force to larger than smaller aggregates (Six et al. 2004) and to moist soil aggregates (Oztas and Fayetorbay 2003). The broken aggregates will expose a greater granular surface area, which can provide more habitats for microbes, and more absorption points and opportunities for dissolved carbon and nitrogen. Freeze–thaw will also affect the microbial biomass and microbial community composition in soils. Freeze–thaw can lead to the death of some microbes because the quickly-formed ice crystals in soil in freezing conditions break microbial cell membranes through physical effects (Macleod and Calcott 1976). Soulides and Allison (1961) and Skogland et al. (1988) found that freeze–thaw could reduce the number of bacteria in soils. Autotrophic

X. Yu, *Material Cycling of Wetland Soils Driven by Freeze–Thaw Effects*,
Springer Theses, DOI: 10.1007/978-3-642-34465-7_1,
© Springer-Verlag Berlin Heidelberg 2013

nitrifying bacteria are particularly sensitive to injury, and take a long time to recover from injury (Focht and Verstraete 1977). When the soil is frozen, the intracellular content of soil microbes is released into the soil environment in the process of the killing of microorganisms. The population of soil microorganisms typically declines by 30–50 % (Macleod and Calcott 1976) and large amounts of nutrients are released following microorganism death. The decline in microorganism populations will lead to lower biological activity by the microbial biomass, which will affect the accumulation, release and cycling of carbon and nitrogen. Current studies of freeze–thaw effects on carbon and nitrogen release in soils are mainly focused on freeze–thaw effects on gas release. For forest ecosystems (Skogland et al. 1988), agriculture ecosystems (Prieme and Christensen 2001), peatland ecosystems in the temperate zone (Bubier 2002), tundra ecosystems (Mikan et al. 2002) and alpine grassland ecosystems (Kato et al. 2005), soil respiration is significantly enhanced after a freeze–thaw period. The amounts of released microbially-produced greenhouse gas released annually, in particular N_2O, are largely controlled by the alternation of soil freeze–thaw and soil dry-moist conditions. Therefore, the melting of permafrost can lead to an increase in released N_2O (Christen and Tiedje 1990; Chang and Hao 2001; Prieme and Christensen 2001). However, most of these studies were limited to indoor simulation tests or field observation experiments, and little attention has been paid to gas emission modeling in wetland soils in large-scale freeze–thaw conditions. Gas emission is only one part of carbon and nitrogen accumulation and release, and the processes of adsorption and desorption of dissolved carbon and nitrogen in soils and the mineralization of organic carbon and nitrogen in soils as well as other processes are also very important.

1.1.2 Significance of the Research

This study focused on understanding the mechanisms by which freeze–thaw affects wetland soil carbon and nitrogen accumulation and release, and to explore the far-reaching environmental effects of the response of wetland soil carbon, nitrogen accumulation and release to freeze–thaw. The results should improve our understanding of the mechanism of wetland material cycles in seasonal freeze–thaw areas. They should also help in understanding the effects of global warming and wetland reclamation on wetland carbon and nitrogen accumulation and release.

1.2 Research Progress: A Worldwide Review

1.2.1 The Effect of Freeze–Thaw on Soil Material Circulation

1.2.1.1 The Soil Freeze–Thaw Process

The term "frozen soil" refers in general to ice-containing soil, and to soil with a temperature at or below 0 °C. When the soil temperature is at or below 0 °C, but does not contain ice, a frozen soil is called a cold soil. In nature, frozen soil layers, or frozen soil areas as a whole, can contain frozen soil together with cold soil.

According to the time for which soils are frozen, frozen soils can be classified into short-time frozen soils (from a few hours, a few days and up to half a month), seasonal frozen soils (from half a month to a few months), and permafrost (from a few years to a few hundred thousands of years).

The near-surface soil layer that freezes every winter and melt every summer is called the active layer. According to the composition of the active layer, there are seasonal melt layers and seasonal frozen layers. The former refers to the seasonal melt on the surface of permafrost, and the latter to the seasonal freeze that occurs on the surface of melted soil.

If the surface of a particular layer is subjected to seasonal freezing and seasonal melting, the corresponding layers are referred to as the seasonal freezing and seasonal melting layers. Seasonal freezing is the freezing of the thawed soil with a mean annual temperature above 0 °C and seasonal melting is the melting of permafrost with an annual average temperature below 0 °C. Freeze–thaw of the permafrost in northeast China is seasonal melt. The seasonal melted layer is the melting layer of permafrost which is near the surface, and is formed through surface heat exchange when the temperature is above zero. This seasonal melt has an important impact on the thermal conditions of the lower part of the soil and production activities.

Many factors can affect seasonal freeze and melt, including snow coverage, vegetation, topography, water bodies, lithological character and water content. These factors actively participate in the heat exchange between the atmosphere and the ground and affect the temperature conditions of the surface and ground, thus determining the seasonal freeze and thaw characteristics of a soil.

The study area is located on the Sanjiang Plain, a seasonal freeze–thaw zone. This region has a long frozen period with the temperature beginning to decline sharply in autumn, and then becoming frozen in late October. It starts to melt in early April of the next year, so the freezing period is more than 6 months and the average freezing depth is 150–210 cm. The frozen layer is generally fully melted by June; but the marsh areas, especially the peat marsh areas, are often not fully melted until July or early August (Ma and Niu 1991) because of the low thermal conductivity of the grass-roots layer or peat layer. The presence of the frozen layer prevents the infiltration of surface water and melted snow and their continued retention on the surface in the long-term forms an over-wet environment, facilitating the formation of marshes.

1.2.1.2 Freeze–Thaw Effects on Soil Physical Properties

Freeze–thaw is an abiotic stress applied to a soil. Freeze–thaw changes the phase changes of soil, and thus soil physical properties change, mainly soil structure, soil permeability and soil mechanical properties. The impact of freeze–thaw on soil physical properties depends on the freeze–thaw rate, freeze–thaw temperature, soil moisture, soil bulk density and the number of freeze–thaw cycles (Edwards 1991; Lehrsch 1998).

The disrupting effect of freeze–thaw on the stability of soil aggregates is affected by the strength of the freeze–thaw process, soil water content, the size of the aggregates and other factors. The more severe the freeze–thaw, the more disruptive it is for soil properties (Fitzhugh 2001). Logsdail and Webber (1959) noted that the alternating contraction and expansion of soil aggregates of lower water stability meant that the damaging effects were enhanced with increased soil water content, and the damaging effects began to decline when the water content exceeded saturation. Oztas and Fayetorbay (2003) considered that freeze–thaw was more disruptive for moist soil aggregates and that this destructive effect was enhanced with an increase in the number of freeze–thaw cycles from three to nine. In terms of the aggregates themselves, freeze–thaw has greater damaging effects on larger aggregates than on smaller ones (Six et al. 2004).

Freeze–thaw can change soil permeability. After a freeze–thaw cycle, although the fine-grained soil void ratio typically decreases, the fine-grained soil permeability increases. Several studies have confirmed that the soil permeability coefficient increases by approximately 1–2 orders of magnitude after freeze–thaw (Chamberlain and Gow 1979). They suggested it was because freeze–thaw of soil facilitates the formation of cracks in the soil and the volume of fine-grained soil in soil pores may reduce during the freeze–thaw process. Later studies have shown that this effect of freeze–thaw can be significantly impeded by pressure (Kim and Daniel 1992). The reduction in void ratio and the increase in permeability which occur in the freeze–thaw process are mainly due to micro-cracks formed during freeze–thaw or to large pores caused by the melting of ice crystals (Chamberlain 1979).

Chamberlain (1979) also found that freeze–thaw often causes a reduction in the void ratio to a certain extent, therefore effectively making it like a freeze–thaw compressed (or compacted) soil. In later studies of soil samples with a greater range of bulk density, it was also found that for loose soil, the freeze–thaw process can reduce the void ratio and thus increase the soil density, but the opposite was found for compacted soil. The dual effects of the freeze–thaw process on the void ratio of soil have gradually become widely accepted. In addition, most studies have found that the soil permeability and bulk density tend to be stable after three to five freeze–thaw cycles (Othman and Benson 1993; Viklander 1998; Boynton and Daniel 1985).

Freeze–thaw can change the mechanical properties of a soil. Graham and Au (1985) carried out a one-dimensional compression test on originally clay textured soil samples before and after freeze–thaw and found that soil structure in the soil was seriously fractured after five freeze–thaw cycles. After studying undisturbed cohesive soil samples after freeze–thaw, Yang and Zhang (2002) and Wang et al. (2001) showed that the unconfined compressive strength and sensitivity were significantly reduced. Although there are more studies on the effect of freeze–thaw on soil strength than on the relationships between stress and strain, the studies are not comprehensive and the results are not systematic. For example, some studies found that the soil strength increased after freeze–thaw, while others found that the soil strength decreased (Wang et al. 1996).

1.2.1.3 The Impact of Freeze–Thaw Effects on Soil Nutrient Circulation

The freeze–thaw process affects nutrient circulation mainly with respect to soil microbial properties, gas release and the nutrient contents in a soil.

Freeze–thaw affects the microbial biomass and community composition in a soil. Freeze–thaw can lead to the death of some microbes because of the physical effects of quickly-formed ice crystals in freezing conditions breaking cell membranes (Macleod and Calcott 1976). Soulides and Allison (1961), and Skogland et al. (1988) found that freeze–thaw will reduce the number of bacteria in a soil. Autotrophic nitrifying bacteria are particularly sensitive to injury and it takes a long time for them to recover from injury (Focht and Verstraete 1977). When soil is frozen, the intracellular contents of soil microbes are released to the soil environment in the process of microbial death and subsequent lysis of the microbial cells. The amounts of nutrients that are released into the soil environment are great, and the population of microorganisms in the soil typically decreases by 30–50 % (Macleod and Calcott 1976). Different microorganisms have very different abilities to resist the effects of freeze–thaw processes. The disturbance from soil freeze–thaw can kill some soil microbes, but stimulate residual microbial activity. These residual microorganisms use dead microbial cells as a matrix, thus increasing their own activity. Factually, it was found a century ago that cold-adapted soil microorganisms could survive and grow below zero degrees and respiration in frozen soils has been repeatedly found in laboratories (Gilichinsky 1995). Ostroumov and Siegert (1996) have shown that when the soil temperature is below freezing, microbial activity is still possible because of a small amount of water in the soil still being unfrozen with the effect that the microorganism matrix and waste can flow and diffuse. Some microorganisms can lower their freezing point through the accumulation of solute, thus reducing the damage caused by freeze–thaw (Focht and Verstraete 1977). Panoff et al. (1998) noted that exposure of some of the moderate temperature microorganisms to freezing soil temperatures, especially freeze–thaw, will result in a slower death rate. The interference of freeze–thaw with soil microorganisms is complicated because of the existence of many relevant factors, such as the speed of the frozen microbial growth phase, the concentration and composition of extracellular solvents, the number of freeze–thaw cycles and the initial soil water content. For some ecosystems, the freeze–thaw process will result in interference with a large part of the microbial biomass (Yanai et al. 2004; Bolter et al. 2005); and for other ecosystems, such as alpine ecosystems and Arctic tundra, a large number of microorganisms can resist interference from freeze–thaw (Lipson and Monson 1998; Lipson et al. 2000; Grogan et al. 2004). In different freeze–thaw cycles, microorganisms respond to freeze–thaw very differently. The population of some microorganisms sharply declined in a small number of freeze–thaw cycles (Walker et al. 2006), while the population of some microorganisms sharply declined after only a few freeze–thaw cycles (Winter et al. 1994).

Freeze–thaw affects soil gas emissions. For forest ecosystems (Skogland et al. 1988), agro-ecosystems (Prieme and Christensen 2001), temperate peat ecosystems (Bubier et al. 2002), tundra ecosystems (Mikan et al. 2002) and alpine grassland ecosystems (Kato et al. 2005), soil respiration has been shown to be significantly enhanced after the freeze–thaw period. The annual emission amounts of greenhouse gases generated by microorganisms, especially N_2O, are largely controlled by the alternation of soil freeze–thaw and soil drying and wetting processes, and therefore the melting of frozen soil can lead to an increase in N_2O emission (Christensen and Tiedje 1990; Chang and Hao 2001; Prieme and Christensen 2001). There are many studies showing that freeze–thaw leads to an increase in N_2O emission. Schimel and Clein (1996) showed that an increase in soil N_2O was due to the release of available nutrients from dead microorganisms into the soil caused by freeze–thaw or the dispersal of soil aggregates after soil melted. Other studies have suggested that nitrification and denitrification are the main reasons for observed increases in N_2O emission after freeze–thaw of soil (Christensen and Tiedje 1990; Beauchamp 1997). When soil freezes, ice crystals are quickly formed, leading to the death of soil microorganisms. However, once the ice crystals have formed they are able to buffer temperature changes to prevent the quick formation of new ice crystals and thus they protect the surviving microorganisms. These living microorganisms can perform nitrification and denitrification and produce N_2O, which accumulates beneath the ice and is not released. However, once the frozen soil melts and the ice destructs, the accumulated N_2O can be released, with a consequent increase in N_2O emission. Rover and Heinemeyer (1998) considered that the increase in N_2O emission was correlated with biological activity. When Sommerfeld et al. (1993) observed N_2O emissions; they also observed CO_2 emission and O_2 absorption, which confirmed Rover and Heinemeyer's suggestion. Studies have shown an increase in N_2O emission with an increased number of freeze–thaw cycles. However, Chen et al. (1995) considered that N_2O emission showed an increasing trend over the first several freeze–thaw cycles, but the emission began to slow in subsequent cycles. In recent years, it has been suggested that not all of the N_2O is released in the melting phase, but that some is released as N_2 because of a reduction in N_2O reductase activity (Koponen and Martikainen 2004). Ludwig et al. (2004) considered that the relationship between N_2O emission and freeze–thaw was correlated with location, and was closely correlated with the dominant plant community.

In contrast to the findings for N_2O release, some studies have indicated that continuous freeze–thaw cycles reduce the emission of CO_2 (Schimel and Clein 1996). Current research on the effect of freeze–thaw on methane emission is not systematic and consistent. Heyer et al. (2002) showed that freeze–thaw can increase emission of methane, while Prieme and Christensen (2001) believed that freeze–thaw did not.

The freeze–thaw process affects soil nutrient contents. Because of the interference of freeze–thaw with litter, soil organic matter and soil microbial activity, some losses of soil carbon and other nutrients occur after the freeze–thaw period (Hobbie and Chapin 1996; Schimel and Clein 1996). Controlled simulation of freeze–thaw cycles led to increased loss of dissolved organic matter (Wang and

Bettany 1994), an increase in mineral nitrogen (Deluca et al. 1992) and a reduction in the effective mineral nitrogen (Sulkava and Huhta 2003). However, mineralization of carbon and nitrogen increased as the number of freeze–thaw cycles increased. The increase rate began to slow down after the third cycle (Schimel and Clein 1996; Herrmann and Witter 2002). The effect of freeze–thaw on soil nitrogen mineralization also varied with study location (Schimel and Clein 1996). When the number of freeze–thaw cycles increased to a certain extent, the impact on carbon and nitrogen tended to be minimal (Grogan et al. 2004). The emission of carbon and nitrogen produced from freeze–thaw has little effect on the carbon and nitrogen balance for the whole year (Herrmann and Witter 2002). Studies have shown that freeze–thaw can increase phosphorus loss (Fitzhugh et al. 2001). Wang et al. (2007) studied the impact of freeze–thaw on adsorption and desorption of phosphorus in wetland soils. He found that freeze–thaw can promote the adsorption and desorption of phosphorus in wetland soil, and that the intensity of soil adsorption and desorption increased with an increase in the number of freeze–thaw cycles from one to seven. The simulation results showed that the total dissolved phosphorus in the soil increased with an increase in the number of freeze–thaw cycles (Vaz et al. 1994). However, the loss of dissolved total phosphorus from freeze–thaw came from plant tissue, rather than soil (Bechmann et al. 2005). Other studies did not find a connection between available phosphorus and freeze–thaw (Peltovuori and Soinne 2005; Soinne and Peltovuori 2005).

1.2.2 Accumulation and Release of Carbon and Nitrogen in Wetland Soils

1.2.2.1 Accumulation and Release of Carbon in Wetland Soils

Carbon Storage in Wetland Soils

Wetlands, which cover about 8–10 % of the global land area, store 10–20 % of the global carbon, and are the largest global carbon reservoir (Dixon and Krankina 1995). Atmospheric CO_2 can be fixed by wetlands and stored in the form of peat. The total amount of carbon in global wetlands is about 770×10^{15} g, and the northern peatlands alone fixed 500×10^{15} g of carbon (Gorham 1991), which is more carbon than is fixed in agro-ecological systems (150×10^{15} g), temperate forest (159×10^{15} g) and tropical rain forest (428×10^{15} g).

Emission of Carbon-Containing Greenhouse Gas

Song et al. (2003, 2004, 2005a, b) studied emission fluxes of CH_4 and CO_2 on the Sanjiang Plain in the growing season and during the freeze–thaw period. They explored the effect of temperature and moisture conditions on gas emission, and compared the characteristics of soil respiration and CH_4 emission before and after reclamation of the Sanjiang Plain marsh wetlands. Wang et al. (2002a, b) compared CH_4 emission and its influencing factors in the Sanjiang Plain wetlands, rice paddy fields, and the Zoige Plateau wetlands. Yang et al. (2004) studied CH_4 emission in the Sanjiang Plain wetland island-shape forest and investigated the influence of temperature and moisture.

Accumulation and Emission of Wetland Soil Organic Carbon

The accumulation of organic carbon in an ecosystem depends on the net primary productivity (NPP) of the vegetation system and the difference between decomposition and net emission of organic carbon. Because of the limitation of wetlands having a high water content and a strong reducing environment, the rate of decomposition and conversion of wetland plant residues is relatively slow and usually organic carbon accumulates as peat or organic matter (Albuquerque and Mozeto 1997). Furthermore, the environmental characteristics of wetland ecosystems have an important impact on the accumulation of organic carbon. For open or semi-open wetland ecosystems, the addition and loss of organic matter along with water and dissolved organic carbon (DOC) account for a certain proportion of the total organic carbon balance. Wetland soil organic carbon mainly exists in the forms of half-decomposed plant residues, half-decomposed products of decomposition, dissolved carbon and humified carbon (Tong et al. 2005). For a particular region, the decomposition of organic carbon is mainly affected by three factors: (1) the nature of the substrate to be decomposed (Belyea 1996); (2) the physicochemical properties of the microenvironment which can affect decomposition, such as temperature, hydrological conditions, pH, and Eh (Donnelly et al. 1990; Berg et al. 1993); (3) the time that the substrate to be decomposed coexists with the decomposing microenvironment. The time that litter is present in the decomposing microenvironment can have a sustained effect.

1.2.2.2 Accumulation and Emission of Nitrogen in Wetland Soils

Nitrogen Transport in Wetland Soils

In China, studies of nitrate nitrogen leaching and its contamination in groundwater began in the early 1980s. In the early 1990s research began to focus on farmland nitrogen leaching (Zhu and Wen 1992). The continuous flooding vertical soil column simulation study of Wang (1997) showed that the transport and transformation processes of nitrate nitrogen in the soil were closely correlated with soil characteristics, and that nitrogen in a fine sand soil was transported faster than in a sandy loam soil. Chen et al. (1995) studied the horizontal transport of nitrate nitrogen in a gleyed paddy soil. They found nitrate nitrogen concentrations gradually decreased with increasing distance from the tracer agent and increased with an increase in soil moisture. They also found that the nitrate nitrogen transport rate varied with transport distance as a power function. Bai (2003) reported that nitrate nitrogen accumulated in the root layer of soil in the Xianghai wetland and nitrate nitrogen leaching below the root zone reduced with increasing soil depth. Until now there are only a few studies of the mechanism of the horizontal transport of nitrate nitrogen. Bai (2003) found that the transport fluxes of nitrate nitrogen and ammonium nitrogen in wetland soils varied with transport distance and followed a first-order exponential decay, but that the movement fluxes of nitrate nitrogen were obviously restrained by diffusivity and soil water content. Compared to closed wetlands, open wetlands are more conducive to nitrogen leaching with vertical nitrogen leaching being affected by soil texture and the mobile water and non-mobile water content in the soil and other factors. Sun (2007) considered that the horizontal transport concentration and rate of wetland soil inorganic nitrogen and distance followed an exponential decay model, and that the movement was controlled by the concentration gradient, water potential gradient, soil matrix potential and adsorption saturation. Breakthrough curves of the vertical transport of inorganic nitrogen are compatible with the Gauss single peak model, but peak shape characteristics of different breakthrough curves for different soil layers varied greatly, being mainly affected by soil clay content, water composition, solute transport mode and nitrification/denitrification.

Decomposition and Transformation of Nitrogen in Wetland Soils

Worldwide studies of decomposition and the transformation of nitrogen in wetland soils mainly focus on N_2O emissions and the mineralization process, as well as nitrification and denitrification processes. N_2O emission is a major route for transporting soil nitrogen to the atmosphere. Liu et al. (2003) in a groundbreaking study, investigated the characteristics of N_2O concentrations and emission in the Sanjiang Plain wetland, which filled a domestic blank in the field of N_2O emissions from natural wetlands. The mineralization and decomposition of soil organic matter might be an input source of soil nitrogen, but it can also be made use of by soil microbes (Liu et al. 2000). The annual amount of net mineralization of wetland soil nitrogen is lower than that in grassland and forest ecosystems, and thus is more conducive to effectively retention of nitrogen. The denitrification activity and rate of the 0–10 cm layer in wetland soils are higher and closely correlated with soil physicochemical properties, and account for between 52.39 and 66.40 % of denitrification nitrogen losses (Sun 2007).

Nitrogen Retention in Wetland Soils

Considerable research has been done in China on the nitrogen purifying capacity of wetlands. Studies of the Baiyangdian wetlands by Yin (1995) showed that the minor groove among water land ecotones can effectively retain land-based nutrients, including retaining up to 42 % of the total nitrogen (TN) that ran off the surface. Qu et al. (2000) reported that a reed wetland in East Liaoning Bay could purify more than 60 % of land-based nitrogen sources. In a study of the purification function of ditch wetlands on the lower reaches region of the Yangtze River for agricultural non-point source pollutants, Jiang et al. (2004, 2005) found that two kinds of natural growth in the ditch, emergent aquatic plants, reed and wild rice effectively absorbed nitrogen; each year 463–515 kg/hm^2 of nitrogen could be removed through harvest, which equates to a nitrogen loss of 2.3–3.2 hm^2 from local farmlands.

1.2.2.3 Adsorption and Desorption of Soil DOC and Ammonium Ions (NH$_4^+$)

Dissolved organic carbon (DOC) is the main controlling factor for soil formation (Dawson et al. 1978), and soil weathering (Raulund-Rasmussen et al. 1998), and it affects soil circulation, microbial activity, organic matter degradation and its transformation (Magill and Aber 2000; Williams et al. 2000). The adsorption of DOC to soil will affect factors important for the transport, fixation and accumulation of organic matter in soil (Kaiser and Zech 2000; Guggenberger and Kaiser 2003). There have been many studies on DOC adsorption in the soil (Kaiser and Zech 2000; Vandenbruwane et al. 2007; Gjettermann et al. 2008).

There are six main mechanisms involved in the absorption of DOC to soil: ligand exchange, cation bridges, anion exchange, cation exchange, van der Waals interactions and hydrophobic effects. DOC adsorption to soil is affected by many factors, such as DOC concentration, soil pH, adsorption time, and soil mineral composition (Moore et al. 1992; Lilienfein et al. 2004; Ussiri and Johnson 2004). The adsorption capacity of DOC to soil is positively correlated with soil clay content, soil solution ionic strength and pH (Shen 1999).

Ammonium (NH$_4^+$) adsorption and desorption processes in soil are important mechanisms underlying the accumulation and release of nitrogen in wetland soils. The nutrients absorbed to wetland soils provide a living environment for anaerobic and aerobic microorganisms so that they can perform nitrification and denitrification and other microbiological activities (Johnston 1991). Adsorption and desorption of soil NH$_4^+$ is related to the plant uptake of available nitrogen and leaching of soil NH$_4^+$ (Avnimelech and Laher 1977; Fenn et al. 1982). Adsorption and desorption of NH$_4^+$ in wetland soils is also correlated with agricultural production and environmental safety. The release of excess NH$_4^+$ from wetland soils will pollute downstream water (Phillips 1999). There are many studies of the adsorption and desorption behavior of NH$_4^+$ in farmland and forest soils (Thompson and Blackmer 1992; Lumbanraja and Evangelou 1994; Wang and Alva 2000). For example, Matschonat and Matzner (1996) studied the main factors that affect adsorption of NH$_4^+$ by forest soils. Some studies have compared the NH$_4^+$ adsorption behavior of different types of soils. The results showed that the adsorption capacity of NH$_4^+$ in sand is less than that in clay and silt (Wang and Alva 2000).

Comparing the results of adsorption and desorption of DOC and NH_4^+ by soil, reveals two common shortcomings. One is that the tested soils are mainly farmland soils and forest soils and hardly any of these studies have investigated adsorption and desorption of DOC and NH_4^+ by wetland soils. However, wetlands are long-term flooded or seasonally flooded ecosystems and adsorption and desorption of DOC and NH_4^+ in wetland soils will not only affect the accumulation and release of DOC and NH_4^+ in soil, it will also affect the concentration of dissolved carbon and nitrogen in the overlying water of wetland soils. Another disadvantage of the earlier studies is that they are all performed at normal temperature, and the adsorption and desorption of DOC and NH_4^+ under freezing and thawing conditions have not been studied. For wetland ecosystems in areas of seasonal freezing and thawing, freeze–thaw is a very common and important environmental factor. Thus the study of the impact of freezing and thawing on the adsorption and desorption of DOC and NH_4^+ in wetland soils can help in the further understanding of carbon and nitrogen circulation in wetland ecosystems of the seasonal freeze–thaw zone.

1.2.3   Development Trends

The preceding literature review shows that wetland soil material circulation is a core content of wetland research; while as an important climate-driven factor of northern wetlands, freeze–thaw processes can affect soil physicochemical properties, and directly or indirectly intervene in carbon and nitrogen circulation in wetlands.

Currently what has been widely studied is the material cycle in wetlands in the growing season and its eco-environmental effects (Lv and Huang 1998; Lv 2002). However, the role of freezing and thawing in the accumulation and release of carbon and nitrogen in wetland ecosystems has not received sufficient attention. Wetland soils are long-term water-saturated, and the accumulation and release of carbon and nitrogen will affect the concentrations of carbon and nitrogen in the overlying water of the upper wetlands and in downstream waters. Sanjiang Plain is located in northeast China and contains China's largest contiguous distribution of freshwater wetlands. The soils there go through a freeze–thaw process every year. Previous studies have shown that the freeze–thaw process has a significant effect on the emissions of N_2O and other gases from wetland soils, but there is a lack of systematic studies on the effect of freeze–thaw on other mechanisms such as the accumulation and release of carbon and nitrogen in the Sanjiang Plain wetland soils, which is constraining in-depth study of the year-round principles which underpin material circulation in wetlands.

After reviewing worldwide research on the impact of freeze–thaw on the biogeochemical cycles of macro elements in wetland soils, it is proposed in this paper to study freeze-thaw driven accumulation and release of typical wetland soil carbon and nitrogen in the Sanjiang Plain. Relevant conclusions have a reference value for the in-depth understanding of the transport and transformation of carbon and nitrogen in wetland soils, and for objectively evaluating the impact of human activities on wetland nutrient circulation.

1.3 Research Topics, Technical Scheme and Innovation

1.3.1 Research Topics

1.3.1.1 Freeze–Thaw of Wetland

During the spring melt period, this study would observe the melting of wetland surface ice and snow and the impact of cover on that melting, as well as the impact of different accumulated water depths on the wetland soil thaw depth and soil temperature. During the spring and autumn freeze–thaw period, I would record temperature changes in the farmland soil layers (5, 10, 15, 25 cm) and temperature changes in the wetland soils with or without grass cover (5, 10, 15, 25 cm).

1.3.1.2 Dynamics of Dissolved Carbon and Nitrogen in Wetland Soils, and in the Overlying Ice Cover and Water in the Freeze–Thaw Period

During the spring melt period, this study would conduct the in situ field simulation of the dynamics of concentration changes of dissolved carbon and nitrogen in wetland soils, and in the overlying ice cover and water.

1.3.1.3 Effect of Freeze–Thaw on DOC Adsorption and Desorption by Wetland Soils

After freeze–thaw treatment, this study would record the change in wetland soil DOC adsorption and desorption by investigating the impact of climate warming on the freeze–thaw cycle, and forecast the impact of climate warming on wetland soil DOC adsorption and desorption.

1.3.1.4 Effect of Freeze–Thaw on NH_4^+ Adsorption and Desorption by Wetland Soils

This study would analyze freeze–thaw effects on the capacity of wetland soils for NH_4^+ adsorption and desorption, explore the physicochemical properties of soil that affect wetland soil NH_4^+ adsorption and desorption, calculate the contribution of adsorption and desorption to NH_4^+ accumulation and circulation in the wetland soils, assess the buffering role of freeze–thaw effects on reclaimed wetland soil NH_4^+, and predict the response of wetland soil NH_4^+ adsorption and desorption to global warming under freeze–thaw effects.

1.3.1.5 The Effect of Freeze–Thaw on Organic Carbon and Organic Nitrogen Mineralization in Wetland Soils

Through an indoor simulation study, this study would investigate the dynamics of mineralization of wetland soil organic carbon and organic nitrogen during the freeze–thaw process and to analyze freeze–thaw effects on the mineralization of soil organic carbon and organic nitrogen.

1.3.1.6 The Dynamic Simulation of the Accumulation and Release of Dissolved Carbon and Nitrogen in Wetland Soil in Freeze–Thaw Conditions

This study would observe the dynamic changes in DOC, NH_4^+–N and NO_3^-–N concentrations in wetland soil solutions from different layers in freeze–thaw and non-freeze–thaw conditions, illuminate the reason for freeze–thaw effects on DOC, NH_4^+–N and NO_3^-–N concentrations in the soil solution, and analyze the spatial heterogeneity of freeze–thaw effects.

1.3.2   Research Methods

1.3.2.1 Sampling Site Selection and Sample Collection

Sampling sites were selected according to the principle of their being "scientific, representative and feasible". The Sanjiang Plain maps, wetland maps and vegetation maps were used as the basis for selection of representative sample sites. At these sites, soil section and water samples were collected, and the experiments were established. Through field surveys, two kinds of marsh wetlands (the third area of the Honghe Farm on the Sanjiang Plain, the Haiwang village) and two kinds of riparian wetlands (Naoli River and Bielahong River) were selected for sampling sites. Each sampling site was divided into three plots according to vegetation type for collecting soil samples.

At each sampling site, the altitude, longitude, latitude, vegetation types and environment surrounding the sampling sites were recorded. Taking into account the research goals of this paper, combined with the common sampling method, the sampling and analytical work was limited to the soil within 40 cm of the wetland surface. Soil sections were collected at 10 cm intervals at each sampling site and the samples were stored in sample bags. One part of the samples was air-dried in the laboratory for routine background analysis of the soil, and the other part of the samples was saved in the form of fresh soil for indoor simulation experiments.

1.3.2.2 Observation of Field Freeze–Thaw Principles

Cover and control experiments were laid out at the selected sampling sites in the Sanjiang Ecological Experimental Station. Melting of the ice and snow on the wetland surface was observed in the spring melt period; the effect of cover on the ice and snow melting was studied; after snow and ice melt, thaw depths in the wetland soils and changes in the soil temperature at different accumulated water depths were observed; temperature changes in the farmland soil layers (5, 10, 15 and 25 cm) in the spring and autumn freeze–thaw period were observed; and temperature changes in the wetland soil layers (5, 10, 15 and 25 cm) with and without grass cover and inside and outside of warm shed in spring and autumn freeze–thaw period were observed.

Soil temperature was measured daily at 8:00 and 14:00 by drilling 5 mm diameter holes 5, 10, 15, and 25 cm long at the sampling points. A Tri-Sense Kit (Model: 37000-95, manufacturer: Cole-Parmer, Vernon Hills, IL, USA) was used to observe soil temperature. The sensitivity of the instrument is ± 0.1 %.

1.3.2.3 Dynamic Simulation of Dissolved Carbon and Nitrogen in Wetland Soil, and in the Overlying Ice and Water During the Freeze–Thaw Period

Two sampling sites (H, M) were selected in the third operational area of the Honghe farms of the Sanjiang Plain. The representative vegetation of the two sites was *Carex lasiocarpa*, and *Calamagrostis angustifolia*, respectively. Detailed information about the sampling sites is presented in Table 1.1.

Soil from 0–10 cm beneath the surface was collected at each sampling point. One part of the soil samples was air-dried and homogenized. Samples were kept in the original state of the soil as much as possible for use in the simulation study. Each layer of the other part of the samples was fully mixed, air-dried, and sieved through a 2 mm sieve to remove most of the fine roots. Samples were stored in airtight plastic bags in the dark for determination of soil physicochemical properties. In the meantime, water overlying the wetland surface was collected at each sampling site, and its physical and chemical parameters were determined.

Table 1.1 Description of the study sites

Site	H	M
Land-cover type	Marsh	Wet meadow
Vegetation type	Sedge (*Carex lasiocarpa*)	*Calamgrostis augustifolia*
Hydrological features	Perennial flooded	Seasonal flooded
Soil type	Humus marsh soil	Meadow marsh soil

Soil samples equivalent to 15 g of dry soil were loaded into 500 ml clean pure water bottles. Thirty bottles were prepared at each sampling site. Then, 50 ml of the collected overlying water from the wetland surface was added to the bottles. The bottles were covered with plastic wraps and a number of holes were poked in the plastic wraps with a fine needle. All bottles were then frozen in a freezer for 24 h at $-15\,^\circ$C.

In the spring freeze–thaw period, a point in the vicinity of each sampling site with sparse vegetation and a well-drained soil was selected. Thirty 15 cm-deep round holes with the same diameter as the pure water bottles were drilled. The bottles containing the soil and overlying water were removed from the freezer and put into the round holes.

On days 1, 2, 4, 6, 8, 16 and 18, three bottles were removed from each sampling site. Ice, water and soil were immediately separated and the DOC, NH_4^+–N and NO_3^-–N concentrations in each were determined. After 18 days, all the samples were removed. DOC in ice and water was analyzed by a TOC analyzer (TOC-V_{CPH}, Shimadzu), and NH_4^+–N and NO_3^-–N were analyzed by a continuous flow analyzer (SAN^{++}CFA, Skalar). DOC, NH_4^+–N and NO_3^-–N concentrations in the soil were analyzed by the same instrument after extracting the soil samples.

1.3.2.4   Wetland Soil DOC Adsorption and Desorption Experiments in Freeze–Thaw Conditions

Three sampling sites were selected within study area, which were named as site H, M and G, respectively (Yu et al. 2010). Sites H and M were wetlands, with predominant plant of *Carex lasiocarpa* and *Calamagrostis angustifolia*, respectively. In order to investigate effects of reclamation on sorption/desorption of DOC, a site of reclaimed wetland (site G) was selected. The soils at site G have been cultivated for 10 years, and the unique crop is soybean. Before reclamation, there was no difference between site G and site M. Specific descriptions of the three sites were shown in Table 1.2.

Soils were collected in the surface layer (0–10 cm). At each of the three sampling sites, three soil samples were collected separately. Plant roots were removed from soil samples. Soil samples were homogenized, air-dried, and sieved over a 2

Table 1.2 Description of the three sampling sites in Sanjiang Plain, Northeast China

Abbreviation			H	M	G
Wetland type			Marsh	Wet meadow	Dryland
Vegetation type			Sedge (*Carex lasiocarpa*)	*Calamgrostis augustifolia*	Soybean (*Glycine max*)
Hydrological features			Perennial flooded	Seasonal flooded	Perennial dry
Soil type	Chinese classification	genetic	Humus marsh soil	Meadow marsh soil	Meadow planosol
	U.S. Classification	Taxonomic	Histic Cryaquepts	Humic Cryaquepts	Aquandic Cryaquepts

mm mesh and then stored in plastic bags with airtight and light-free at 20 °C until used in the chemical analysis and sorption experiment. The soil analytical methods are presented at the end of this section.

To simulate the field condition, the peat water extract was used as DOC stock solution for the sorption experiments. The surface peat (0–10 cm) was collected from a site in north of site H. It was away from site H about 200 m. This site was perennial flooded and anaerobic. Thus it was abundant in peat. The peat sample was filled in a glass bottle. Deionized water was added with water peat ratio of 10:1. Then the capped glass bottle was placed on a shaker (130 rpm) for 4 h. The peat water extract was filtered through 0.45 μm polyethersulfone membrane disc filters. The filtrate was concentrated in order to obtain the DOC stock solution with high concentration, which was diluted to five DOC concentrations between 0 and 600 mg/L. It is true that we have not observed so high concentration of DOC in the environment at present. DOC content in soils of the study area was 197–2762 mg/kg. However, the water supply of some wetlands in Sanjiang Plain is decreasing, which causes the accumulation of DOC in soils and the soil surface water. Considering the extreme circumstance, we added the high concentration of DOC in the sorption experiment. Moreover, the high concentration of DOC added might help us to seek the upper limit of the linear increment of DOC sorption.

For the sorption/desorption experiment, each of the three soils was treated by freeze–thaw with 3 replicates as described below. Each soil sample had a control that was untreated by freeze–thaw. Soils with and without freeze–thaw treatment were named as FTT soils and UT soils in the following text, respectively. The sorption/desorption data of DOC were obtained through sequential steps in the following.

Sorption Procedure

Five grams of soils were placed in the 250-ml pre-weighed centrifuge tubes. To each sample, the 50 ml of solution containing varied amounts of DOC (0, 50, 100, 200, 400, and 600 mg/L) was added, which correspond to 0, 500, 1000, 2000, 4000 and 6000 mg/kg DOC added, respectively. Capped tubes were placed on a shaker (130 rpm) at +5 °C for 1 h, and then were frozen at −15 °C for 10 h, and finally thawed at +5 °C for 13 h. This is one freeze–thaw cycle (FTC). The control tubes were stored at 5 °C for one 23 h cycle corresponding to one cycle. The tubes which have been subjected to one to five cycles, respectively, were centrifuged at 5000 rpm for 20 min. Then the supernatant was taken and DOC in the extracts was determined. Adsorbed DOC with one to five cycles was calculated as the difference between the DOC added and DOC remained in the equilibrating solution.

Desorption Procedure

Two DOC samples (4000 and 6000 mg/kg DOC added in the sorption experiment) after 5 cycles were selected, the supernatant was removed after centrifugation and the tubes were re-weighed to determine the volume of solution entrapped in the residue. Then deionized water was added to bring the total volume to 50 ml. After the first cycle as described in sorption procedure, DOC in the supernatant was determined. The difference between the DOC in the supernatant and the DOC in the residue solution was the desorption amount of DOC. Following the first step again, the amount of DOC desorbed by soils after the second to the fifth cycle was calculated. The flowchart of the sorption/desorption experiment procedures of DOC was shown in Fig. 1.1.

Clay content of soils was determined using a Mastersizer 2000 Laser Grainsize (Manufactured by Malvern Instruments Ltd. UK; measuring range: 0.02–2000 μm)

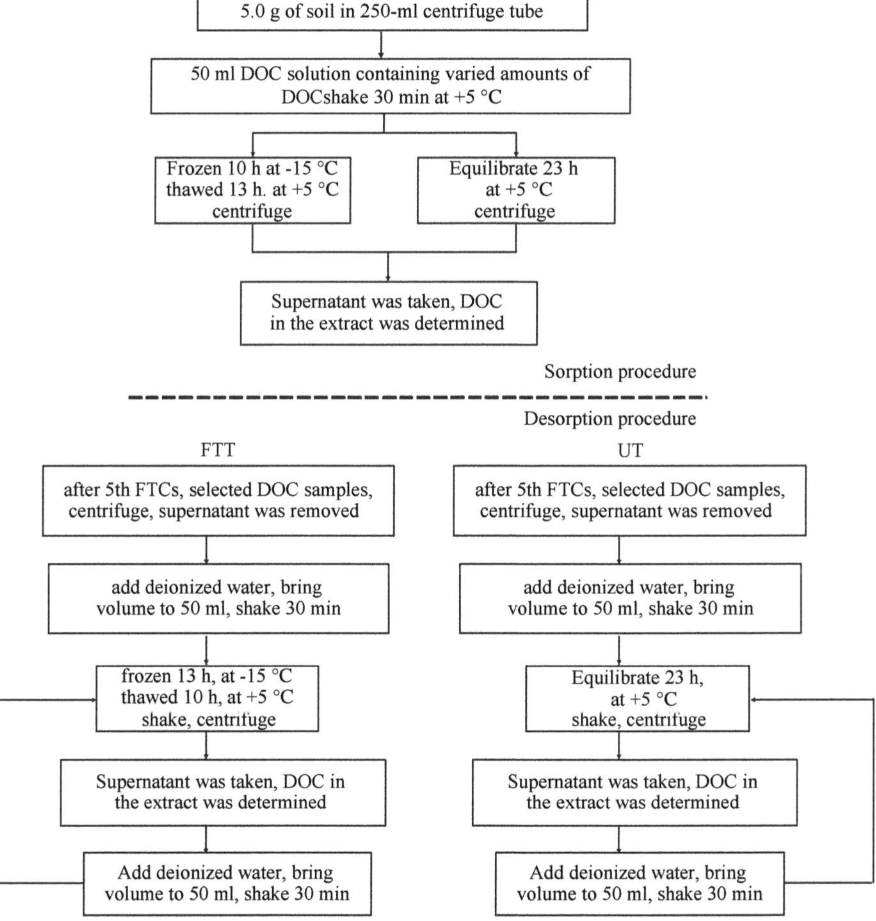

Fig. 1.1 Flowchart of the experiment procedures of dissolved organic carbon (DOC) sorption/ desorption in wetland soils. FTT represents freeze–thaw treatment; UT represents without control treatment (with kind permission from Springer Science+Business Media: Yu et al. 2010, Fig. 1, and any original (first) copyright notice displayed with material)

as described by Wang et al. (2006). Organic matter content of soils was measured by potassium dichromate volumetric method—external heating (Bao 2000). Soil DOC in the extract was determined with TOC-V$_{CPH}$ (Shimazu, Japan). The pH was determined in situ with composite electrodes. The NO_3^-, NH_4^+, and PO_4^{3-} in stock solution were determined using SAN^{++} Continuous Flow Analyzer (SKALAR, Netherland). The supernatant samples were analyzed for DOC concentration. The DOC was determined by SAN^{++} Continuous Flow Analyzer (SKALAR, Netherland).

1.3.2.5 Wetland Soil NH_4^+ Adsorption and Desorption Experiments in Freeze–Thaw Conditions

Three areas were selected for soil sampling in the Sanjiang Plain. The first area was located in the bank of the Naoli River (Area R) (Yu et al. 2011a). In Area R, the sites along the water gradient with the 3 main plant communities *Carex lasiocarpa–Carex pseudocuraica*, *Calamagrostis angustifolia*, and *Oryza sativa* in the area were selected and name R-1 (riverine wetland), R-2 (riverine wetland), and R-3 (paddy field), respectively. The second area was located in the Sanjiang Nature Reserve (Area DF), near the Ussuri River; the sites with the main communities *Carex lasiocarpa–Carex pseudocuraica*, *Calamagrostis angustifolia*, and *Glycine max* in the area were selected and named DF-1 (riverine wetland), DF-2 (riverine wetland), and DF-3 (dry farmland), respectively. The third area was located in the southeast of the Sanjiang Mire Wetland Experimental Station, Chinese Academy of Sciences (Area DI).

In Area DI, there were also three sites selected with the main three communities *Carex lasiocarpa–Carex pseudocuraica*, *Calamagrostis angustifolia* and *Glycine max* in this area, named DI-1 (palustrine wetland), DI-2 (palustrine wetland), and DI-3 (dry farmland), respectively. The description of the nine sites was given in Table 1.3. The farmland sites R-3, DF-3, and DI-3, all converted from natural wetlands 10–20 years ago, were selected in order to investigate effect of wetland reclamation on sorption and desorption of NH_4^+. At each site, three soil samples of the surface layer (0–10 cm) were collected. Each soil sample was air-dried, homogenized, sieved through a 2-mm mesh to remove most fine roots, and then stored in air-tight and opaque plastic bags at 20 °C until use for soil chemical analyses and sorption and desorption experiments.

For the sorption experiment, soil samples 5 g each were weighed after oven-dried and placed in 250-mL pre-weighed centrifuge tubes in 3 replicates. 50 mL of the solutions containing 0, 39, 77, and 154 mg/L (0, 390, 770, and 1,540 mg/kg) NH_4^+ were added to the tubes. Two drops of chloroform were added to inactivate microorganisms. The tubes were capped and placed on a shaker at 130 rpm at 5 °C for 1 h. Then, half of them (FTT) were frozen at −15 °C for 10 h and afterwards were thawed at 5 °C for 13 h, which was set as an FTC; the other half tubes (UT) were stored at 5 °C for a 23 h cycle corresponding to one FTC. Subsequently, all tubes having been subject to 1–5 cycles were divided into three groups: the 1st, 3rd, and 5th FTCs. After each cycle, all tubes were taken out and centrifuged at

Table 1.3 Description of the sampling sites in the study

Plot	Site	Location	Wetland type	Hydroperiod	Dominant plant	Soil type
R	R-1	47°15.937′N 133°45.729′E	Riverine wetland	Perennial water logging	Sedge (*Carex lasiocarpa* and *Carex pseudocuraica*)	Humus marsh soil
	R-2	47°15.844′N 133°45.841′E	Riverine wetland	Seasonal water logging	Narrow-leaf small reed (*Calamagrostis angustifolia*)	Meadow marsh soil
	R-3	47°16.791′N 133°46.095′E	Paddy field	Seasonal water logging	Paddy rice (*Oryza sativa*)	Gley albic soil
DF	DF-1	47°46.079′N 134°38.966′E	Riverine wetland	Perennial water logging	Sedge (*Carex lasiocarpa* and *Carex pseudocuraica*)	Humus marsh soil
	DF-2	47°46.113′N 134°38.979′E	Riverine wetland	Seasonal water logging	Narrow-leaf small reed (*Calamagrostis angustifolia*)	Meadow marsh soil
	DF-3	47°46.115′N 134°38.853′E	Dry land	Dry	Soybean (*Glycine max*)	Meadow albic soil
DI	DI-1	47°34.746′N 133°29.535′E	Palustrine wetland	Perennial water logging	Sedge (*Carex lasiocarpa* and *Carex pseudocuraica*)	Humus marsh soil
	DI-2	47°34.639′N 133°29.569′E	Palustrine wetland	Seasonal water logging	Narrow-leaf small reed (*Calamagrostis angustifolia*)	Meadow marsh soil
	DI-3	47°31.986′N 133°53.019′E	Dry land	Dry	Soybean (*Glycine max*)	Meadow albic soil

R, DF and DI represent Naoli River riverine wetland, Ussuri River riverine wetland and Sanjiang Station palustrine wetland, respectively. R-1. R-2 and R-3 represent three plant communities within plot R. The same is true for plot DF and DI

5000 rpm for 20 min. Then the supernatant was taken and NH_4^+ in the extract was determined. Differences between the amount of NH_4^+ added and that remaining in the equilibrating solution was calculated as the amount of sorption.

For the desorption experiment, the FTT tubes and control tubes with 770 and 1,540 mg/kg NH_4^+ added in the sorption procedure, which had been subjected to five FTCs in the sorption experiment, were selected. After the supernatant was removed through centrifugation in the sorption experiment, these tubes were weighed to determine the volume of solution entrapped in the residue. Then a KCl solution (0.01 mol/L^{-1} KCl) was added to reach the total volume of 50 mL. The FTT and UT tubes that were subjected to one FTC were centrifuged at 5,000 r min^{-1} for 20 min and the supernatant was determined for NH_4^+ concentration. The difference between the amount of NH_4^+ in the supernatant and the amount in the residual solution was the desorption amount. After the first desorption, the supernatant in the tubes was removed and the residue was used for subsequent six successive desorptions, repeating the 1st desorption protocol described above. The amounts of NH_4^+ desorbed were obtained only after the third, fifth, and seventh desorptions.

Soil texture was determined by the pipette method (Gee and Bauder 1986). Soil total organic carbon (C_{org}) was measured by the potassium dichromate volumetric method with external heating (Bao 2000). Soil NH_4^+ extracted by KCl solution and NH_4^+ concentrations in the supernatants were determined with the SAN^{++} Continuous Flow Analyzer (SKALAR, The Netherlands). Soil K^+ and Ca^{2+} were measured by ICPS-7500 (Shimadzu, Japan). The soil cation exchange capacity (CEC) was measured following the method described by Bao (2000).

1.3.2.6 Wetland Soil Organic Carbon and Organic Nitrogen Mineralization Experiments in Freeze–Thaw Conditions

The soils (H, M, G) used in the wetland soil organic carbon and organic nitrogen mineralization experiments in freeze–thaw conditions were the same soils used in the DOC adsorption and desorption experiments. Detailed information about the sampling is presented in Table 1.4.

Soils from the 0–15 cm layer were collected at each sampling point. Grass roots and other debris were removed. Part of the air-dried soil samples was used for the analysis of soil organic carbon (SOC), total nitrogen (TN), nitrate nitrogen (NO_3^-–N), ammonium nitrogen (NH_4^+–N) and other physical and chemical

Table 1.4 Description of the study sites

Site	H	M	G
Land-cover type	Marsh	Wet meadow	Dryland
Vegetation type	*Carex lasiocarpa*	*Calamgrostis augustifolia*	*Glycine max*
Hydrological features	Perennial flooded	Seasonal flooded	Perennial dry
Soil type	Humus marsh soil	Meadow marsh soil	Meadow planosol

parameters. The remaining samples were placed in 50 cm diameter plastic buckets, distilled water was added, and then mixed thoroughly to ensure that the soil moisture content of each sample was 50 %. The well-mixed samples were sealed and incubated in a 25 °C incubator for 7 days. One hundred grams of cultured samples was put into a 1 L wide-mouth flask so that the soils lay evenly on the bottom of the flask. A small beaker containing 10 mL of 0.02 M NaOH solution was put into the flask. Vaseline was applied evenly at the mouth of the wide-mouth bottle, and then the cap was tightened to seal the flask. Thirty-six parallel samples were prepared at each sampling site. Samples were processed in the following two groups: I. The freeze–thaw treatment: samples were completely frozen at -15 °C for 10 h, and then completely melted for 14 h at 5 °C. This treatment was considered to represent a freeze–thaw cycle. II. Control treatment: samples were stored at -5 °C. 24 h represented a cycle. Three triplicate samples were removed from both of the freeze–thaw treatment group and control group at days 1, 2, 5, 9, 14 and 26, representing the 36 samples from each sampling point. The beaker containing NaOH was removed from the flask and 1 M $BaCl_2$ solution was added to it, with the remaining NaOH titrated with 0.05 M HCl. Two blanks were processed for each group. The soil samples were removed for the analysis of TN, NO_3^--N and NH_4^+-N contents. The difference in concentrations of NO_3^--N and NH_4^+-N before and after incubation represents the amount of mineralization of the soil organic nitrogen. TN was determined by the Kjeldahl method. The soils were extracted with KCl solution, and then NO_3^--N and NH_4^+-N concentrations were analyzed on a continuous flow analyzer ($SAN^{++}CFA$, Skalar). Soil physical and chemical parameters were analyzed as previously described.

CO_2–C emission was calculated as follows:

$$C O_2 - C = \frac{(V_0 - V) \times C_{HCl}}{2} \times 44 \times \frac{12}{44} \times \frac{1}{m \times (1 - a\%)}$$

where CO_2–C is the mineralization emission of soil organic carbon during the incubation period, V_0 is the volume of standard hydrochloric acid consumed in the blank calibration, V is the volume of standard hydrochloric acid consumed in the sample titration, C_{HCi} is the standard concentration of the hydrochloric acid, M is the weight of soil sample used, and a % is the soil moisture content.

1.3.2.7 The Dynamic Simulation of Accumulation and Release of Wetland Soil Dissolved Carbon and Nitrogen in Freeze–Thaw Conditions

Four sampling sites were selected in the Sanjiang Plain: Three of the sites were sedge meadows or wet grasslands (*Carex* marsh, *Carex* marshy meadow, and *Calamagrostis* wet grassland) dominated by *Carex lasiocarpa*, *Carex meyeriana* and *Calamagrostis angustifolia*, respectively (Yu et al. 2011b). The fourth site was a soybean field (*Glycine max*) reclaimed 10 years ago from a wetland dominated by *Carex lasiocarpa*. The soybean field is low-lying. There is very deep snow in

Table 1.5 Description of sampling sites in the Sanjiang Plain, northeast China

Site	*Carex* marsh	*Carex* marshy meadow	*Calamagrostis* wet grassland	Soybean field
GPS coordinates	47°35.412′N 133°38.698′E	47°35.397′N 133°38.694′E	47°35.367′N 133°38.663′E	47°35.282′N 133°38.639′E
Wetland type	*Carex* marsh	*Carex* marshy meadow	*Calamagrostis* wet grassland	Soybean field
Vegetation type	*Carex lasiocarpa* Ehrh. community	*Carex meyeriana* Kunth community	*Calamagrostis angustifolia* Kom. community	*Glycine max* (L.) Merr.
Hydrological features	Perennial flooded	Seasonal flooded	Seasonal dry	Perennial dry but flooded in spring freeze–thaw season
Soil type	Histic Cryaquepts	Humic Cryaquepts	Humic Cryaquepts	Aquandic Cryaquepts

the winter. The melting snow produced considerable snowmelt in the next spring, making the soybean field to be water-logged in the spring. Specific descriptions of the four sites were shown in Table 1.5. The living vegetation was removed on top of the sampled soil columns. Soils of the 0–10 cm, 10–20 cm, 20–30 cm and 30–40 cm layers were collected with three replicates and were analyzed for clay content, total organic carbon, DOC content, NH_4^+-N content, NO_3^--N content and phosphorus content.

The replicate soils derived from each site were transported to the lab. We spent about 1 day transporting the soils to the lab. Then we mixed them gently so as not to disturb the soil structure and used to fill six Polypropylene columns (diameter 10 cm, length 50 cm) layer by layer. There was no living vegetation on top of the columns. However, there were a lot of dead or dying roots present in wetland soils, especially the upper three soil layers. Also some dead or dying roots were present in the 0–10 cm soil layer of the soybean field. All the roots were not taken out in the experiment, in order to hold the original condition of the soils. The disturbance in such a sample can be expected to be smaller than in many other experiments where samples were taken from different depths, dried, sieved and well mixed. However, the term "undisturbed soil column" suggests the sampling of a soil monolith that is transferred to the laboratory in one piece for further experiments, like done in many other studies (Teepe et al. 2001; Goldberg et al. 2008; Hentschel et al. 2008). Thus the soil columns in this study were defined as semi-disturbed soil columns. The sidewall and the bottom of the columns were wrapped by thermal insulation materials to impose freezing from the top down (Fig. 1.2). Then the soil solution extractors were installed in each soil layer. The soil solution extractor (designed by Institute of Soil Science, Chinese Academy of Sciences) is composed of the organic glass tube and the porous ceramic filter with a pore size of 0.2–0.5 μm. When the air in the tube was pulled out by an injector, it could utilize

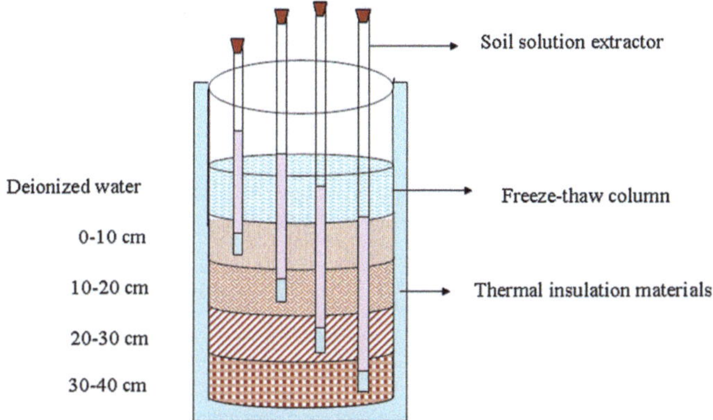

Fig. 1.2 Laboratory experimental device diagrams. The device contains freeze–thaw column, soil solution extractor and thermal insulation materials. The soil solution extractor is composed of the organic glass tube and the porous ceramic filter with a pore size of 0.2–0.5 μm. The soil solution extractor can utilize negative pressure to extract the soil solution. The thermal insulation materials can guarantee that the interchange of heat mainly appears on the soil surface, which is consistent with the natural condition (reprinted from Yu et al. 2011b, with permission from Elsevier)

negative pressure to extract the soil solution in vacuum tubes through the porous ceramic filter. In order to avoid the edge effect, the soil solution extractor was permeable only at the tip. Thus it worked only in a well-defined depth (about 5 cm) around maximum insertion depth. The advantages of this soil solution sampling method are convenient, timesaving and without sampling soils from soil columns and destroying soil structure (Xin et al. 2005). However, the dissolved components might not be extracted completely because of the limited negative pressure. Thus the measurable values might be a little lower than that extracted by K_2SO_4 solution (Xin et al. 2005).

Before the experiment, all columns were filled with deionized water until the water level kept 10 cm above the soil surface. All flooded columns were stored at a constant temperature of 5 °C for 14 days. Then three soil columns (replicates) of each site were transferred to the freezing incubator, which was maintained at -10 °C. The other three columns remained unfrozen in the constant temperature incubator and served as controls. After 1 day, the columns in the freezing incubator were transferred back to the constant temperature incubator and incubated at 5 °C for another 7 days together with the control columns. Thus, in total, one freeze–thaw cycle lasted 8 days. We carried out a pilot test before the actual experiment to investigate how long it takes for the soil columns to completely freeze and thaw again. We found that the flooded soil columns could be frozen completely at -10 °C after 1 day, and then be thawed completely under the condition of 5 °C after 7 days. The extent of freezing and thawing was detected directly by using a steel stick to pierce soil columns. Each column was subjected to seven cycles in the experiment. After each cycle, the solution in each soil layer was collected using the soil solution extractor, and DOC, NH_4^+–N, NO_3^-–N and TDP were determined.

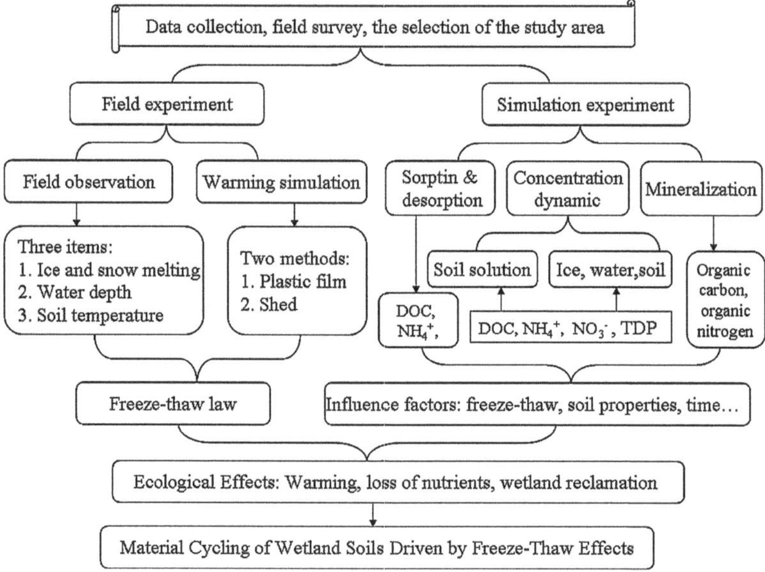

Fig. 1.3 Technical sketch of the research

Clay content of the soil was determined using a Mastersizer 2000 Laser Grainsize (Manufactured by Malvern Instruments Ltd. UK; measuring range: 0.02–2000 μm) as described by Wang et al. (2006). Soil organic carbon was measured by potassium dichromate volumetric method requiring external heating. DOC was determined with TOC-V$_{CPH}$ (Shimazu, Japan). NH_4^+–N, NO_3^-–N, phosphorus and TDP were determined using SAN^{++} Continuous Flow Analyzer (SKALAR, The Netherlands).

1.3.3 Technical Scheme

See Fig. 1.3.

1.3.4 The Key Issues Addressed

Clarification of the response mechanism of accumulation and release of the wetland soil carbon and nitrogen to freeze–thaw effects.

Exploration of the effect of freeze–thaw on the far-reaching environmental effects of accumulation and release of wetland soil carbon and nitrogen.

1.3.5 Innovation

Climatic conditions are the main determinant of the ecosystem condition of a region. For many years, research has focused on the effect of water, heat and other climatic factors on ecosystems, whereas research on the effects of the freeze–thaw process on ecosystems is relatively rare. Freeze–thaw is an abiotic stress acting on a soil and it is a common natural process in medium and high latitudes. In recent years, although some studies have investigated the effect of freeze–thaw effects on soil physical and chemical properties, most have focused on agricultural soils or forest soils, while little attention has been paid to the effects of freeze–thaw on the accumulation and release of wetland soil carbon and nitrogen system, which is the main research aims of this paper.

The main innovations of this paper are as follows:

I   Dynamic changes in the concentrations of dissolved carbon and nitrogen in wetlands soil in the freeze–thaw period can reflect the effects of the freeze–thaw process on the accumulation and release of wetland soil carbon and nitrogen. There have been many studies in recent years on the effects of freeze–thaw on soil dissolved carbon and nitrogen concentrations. However, the results vary widely and some of the conclusions are contradictory, mainly because of different experimental methods. In particular, the temperature and time of freeze–thaw and the season in which the soil samples are collected can have a large impact on the experimental results. In this study, through a semi-simulation experiment in a field micro-area, making use of field freeze–thaw temperatures and time in the spring freeze–thaw period, we studied the dynamic changes in DOC and NH_4^+–N and NO_3^-–N concentrations in wetland soil, and in the overlying ice and water, which truly reflect the process of wetland soil carbon and nitrogen accumulation and release in the freeze–thaw period.

II   Current research on freeze–thaw effects on soil material circulation mainly focuses on biochemical processes, and little attention has been paid to the research of the freeze–thaw effects on soil physical and chemical processes, such as adsorption and desorption. In fact, the adsorption of DOC to soil not only controls the element transfer processes of organic matter and auxiliary organic matter, but it also controls the stability and accumulation of organic matter in soil. Wetland soil NH_4^+ adsorption is the main nitrogen retention mechanism. By comparing the adsorption and desorption behavior of DOC and of NH_4^+ in the freeze–thaw treated wetland and the control wetland soil, it is found that freeze–thaw may improve the adsorption capacity of DOC and NH_4^+ by wetland soil. Reducing its desorption capacity is a key step through which the freeze–thaw process affects the accumulation and release processes of wetland soil carbon and nitrogen. This further clarifies the main mechanism through which freeze–thaw affects soil biogeochemical processes in wetlands. This provides a theoretical foundation for understanding the characteristics of wetland material circulation in seasonal freeze–thaw regions.

III   Chemical changes in the soil solution can reflect changes in the soil system in real time. In almost all previous research, the K_2SO_4 extraction method was used to study soil-dissolved nutrients in freeze–thaw conditions. But this method will change the chemical composition of the soil solution and affect ion balance, making it difficult to reveal the real condition of the soil solution in the field. Therefore, it is necessary to use an in situ non-destructive soil solution collector. This study developed an in situ soil column simulation device, which was suitable for freeze–thaw experiments and which could continuously non-destructively collect soil solution from different soil layers. This paper also presents data from a systematic study of the effects of freeze–thaw treatment on DOC, NH_4^+–N and NO_3^-–N concentrations in wetland soils and agricultural soils, and analyzes the spatial heterogeneity of freeze–thaw effects. Research methods in this paper are innovative.

In summary, the findings of this paper will help the further study of wetland biogeochemical processes in seasonal freeze–thaw regions, and contribute to an understanding of the effects of global warming on the accumulation and release of wetland carbon and nitrogen.

1.4 Chapter Summary

This chapter discusses the research background of the paper, and the purpose and significance of the research. After comprehensively reviewing worldwide studies of freeze–thaw effects and wetland material circulation, we proposed the study of freeze–thaw effects on the accumulation and release of soil carbon and nitrogen in different types of wetland soil. We decided to study four types of soil, a perennially flooded *Carex lasiocarpa* marsh and a seasonally flooded *Calamagrostis angustifolia* marsh, a seasonally droughty *Calamagrostis angustifolia* marsh meadow, artificially seasonally flooded rice fields and seasonally submerged drained dry land. Through in situ observations and control experiments, the principles underlying the effect of freeze–thaw driven accumulation and release of wetland soil carbon and nitrogen and its influencing factors were revealed.

References

Albuquerque ALS, Mozeto AA. C:N:P ratios and stable carbon isotope compositions as indicators of organic matter sources in a riverine wetland system (Moji Guaeu River, Sao Paulo, Brazil). Wetlands. 1997;17:1–9.

Avnimelech Y, Laher M. Ammonium volatilization from soils: equilibrium considerations. Soil Sci Soc Am J. 1977;41:1080–4.

Bai J. Study on the biogeochemical processes of nitrogen in Xianghai mire wetlands. Graduate University of Chinese Academy of Sciences. 2003 (in Chinese).

Bao SD. Soil agrochemical analysis. 3rd ed. Beijing: Agriculture Press; 2000 (in Chinese).

Beauchamp EG. Nitrous oxide emission from agriculture soils. Can J Soil Sci. 1997;77:113–23.

Bechmann ME, Kleinman PJA, Sharpley AN, Saporito LS. Freeze–thaw effects on phosphorus loss in runoff from manured and catch–cropped soils. J Environ Qual. 2005;34:2301–9.

Belyea LR. Separating the effects of litter quality and microenvironment on decomposition rates in a patterned peatland. Oikos. 1996;77:529–39.

Berg B, Berg MP, Bottner P. Litter mass loss rates in pine forests of Europe and eastern United States: some relationships with elimate and litter quality. Biogeochemistry. 1993;20:127–59.

Bolter M, Soethe N, Horn R, Uhlig C. Seasonal development of microbial activity in soils of northern Norway. Pedosphere. 2005;15:716–27.

Boynton SS, Daniel DE. Hydraulic conductivity tests on compacted clay. J Geotech Eng. 1985;116:1549–67.

Bubier J, Crill P, Mosedale A. Net ecosystem CO2 exchange measured by autochambers during the snow–covered season at a temperate peatland. Hydrol Process. 2002;16:3667–82 (special issue).

Bullock MS, Kemper WD, Nelson SD. Soil cohesion as affected by freezing, water content, time and tillage. Soil Sci Soc Am J. 1988;52:770–6.

Chamberlain EJ, Gow AJ. Effect of freezing and thawing on the permeability and structure of soils. Eng Geol. 1979;13:73–92.

Chang C, Hao X. Source of N2O emission from a soil during freezing and thawing. Phyton Ann Rei Botanicae. 2001;41:49–60.

Chen Y, Tessier S, Mackensie AF, Laverdiere MR. Nitrousoxide emission from an agricultural soil subjected to different freeze–thaw cycles. Agric Ecosyst Environ. 1995;55:123–8.

Christensen S, Tiedje JM. Brief and vigorous N2O production by soil at spring fall. J Soil Sci. 1990;41:1–4.

Dawson HJ, Ugolini FC, Hrutfiord BF, Zachara J. Role of soluble organics in the soil processes of a podzol, Central Cascades, Washington. Soil Sci. 1978;126:290–6.

DeLuca TH, Keeney DR, McCarty GW. Effect of freeze–thaw events on mineralization of soil nitrogen. Biol Fertil Soils. 1992;14:116–20.

Dixon RK, Krankina ON. Can the terrestrial biosphere be managed to conserve and sequester carbon? Carbon sequestration in the biosphere: processes and products. NATO ASI Ser I Glob Environ Change. 1995;33:153–79.

Donnelly PK, Entry JE, Crawford DI. Cellutose and ligmin degradation in forest soils: response to moisture, temperature, and acidity. Microb Ecol. 1990;20:289–95.

Edwards LM. The effect of alternate freezing and thawing on aggregate stability and aggregate size distribution of some Prince Edward Island soil. J Soil Sci. 1991;42:193–204.

Fenn LB, Datcha JE, Wo E. Substitution of ammonium and potassium for added calcium in reduction of ammonia loss from surfaced-applied urea. Soil Sci Soc Am J. 1982;46:771–6.

Fitzhugh RD, Driscoll CT, Groffman PM, Tierney GL, Fahey TJ, Hardy JP. Effects of soil freezing, disturbance on soil solution nitrogen, phosphorus, and carbon chemistry in a northern hardwood ecosystem. Biogeochemistry. 2001;56:215–38.

Focht DD, Verstraete W. Biochemical ecology of nitrification and denitrification. Adv Microb Ecol. 1977;1:135–214.

Gee GW, Bauder JW. Particle-size analysis. In: Klute A, editor. Methods of soil analysis. Part I: Physical and mineralogical methods. 2nd ed. Madison: American Society of Agronomy, Crop Science Society of America and Soil Science Society of America; 1986.

Gilichinsky D, Wagener S. Microbial life in permafrost: a historical review. Permafr Periglac Process. 1995;6(3):243–50.

Gjettermann B, Styczen M, Hansen HCB, Vinther FP, Hansen S. Challenges in modeling dissolved organic matter dynamics in agricultural soil using. Soil Biol Biochem. 2008;40(6):1506–18.

Goldberg S, Muhr J, Borken W, Gebauer G. Fluxes of climate-relevant trace gases between a Norway spruce forest soil and the atmosphere during repeated freeze-thaw cycles in mesocosms. J Plant Nutr Soil Sci. 2008;171:729–39.

Gorham E. Northern peatlands: role in the carbon cycle and probable responses to climatic warming. Ecol Appl. 1991;1:182–95.

Graham J, Au VCS. Effects of freeze–thaw and softening on a natural clay at low stresses. Can Geotech J. 1985;22:69–78.

Grogan P, Michelsen A, Ambus P, Jonasson S. Freeze–thaw regime effects on carbon and nitrogen dynamics in sub-arctic heath tundra mesocosms. Soil Biol Biochem. 2004;36:641–54.

Guggenberger G, Kaiser L. Dissolved organic matter in soil: challenging the paradigm of sorptive preservation. Geoderma. 2003;113:293–310.

Hentschel K, Borken W, Matzner E. Repeated freeze–thaw events affect leaching losses of nitrogen and dissolved organic matter in a forest soil. J Plant Nutr Soil Sci. 2008;171:699–706.

Herrmann A, Witter E. Sources of C and N contributing to the flush in mineralization upon freeze–thaw cycles in soils. Soil Biol Biochem. 2002;34:1495–505.

Heyer J, Berger U, Kuzin IL, Yakovlev ON. Methane emissions from different ecosystem structures of the subarctic tundra in Western Siberia during midsummer and during the thawing period. Tellus Ser B Chem Phys Meteorol. 2002;54:231–49.

Hobbie SE, Chapin FS. Winter regulation of tundra litter carbon and nitrogen dynamics. Biogeochemistry. 1996;35:327–38.

Jiang C, Cui G, Fan X, Zhang Y. Accumulation of non-point source pollutants in ditch wetland and their uptake and purification by plants. Chin J Appl Ecol. 2005;16(7):1351–4 (in Chinese).

Jiang C, Cui G, Fan X, Zhang Y. Purification capacity of ditch wetland to agricultural non-point polutants. Environ Sci. 2004;25(2):125–8 (in Chinese).

Johnston CA. Sediment and nutrient retention by freshwater wetlands: effects on surface water quality. Crit Rev Environ Control. 1991;21:491–565.

Kaiser K, Zech W. Sorption of dissolved organic nitrogen by acid subsoil horizons and individual mineral phases. Eur J Soil Sci. 2000;51:403–11.

Kato T, Hirota M, Tang Y, Cui X, Li Y, Zhao X, Oikawa T. Strong temperature dependence and no moss photosynthesis in winter CO2 flux for a Kobresia meadow on the Qinghai–Tibetan plateau. Soil Biol Biochem. 2005;37:1966–9.

Kim WH, Daniel DE. Effects of freezing on hydraulic conductivity of compacted clay. J Geotech Eng. 1992;118:1083–97.

Koponen HT, Martikainen PJ. Soil water content and freezing temperature affect freeze–thaw related N2O production in organic soil. Nutr Cycl Agroecosyst. 2004;69:213–9.

Lehrsch GA. Freeze–thaw cycles increase near–surface aggregate stability. Soil Sci. 1998;163:63–70.

Lilienfein J, Qualls RG, Uselman SM, Bridgham SD. Adsorption of dissolved organic and inorganic phosphorus in soils of a weathering chronosequence. Soil Sci Soc Am J. 2004;68:620–8.

Lipson DA, Monson RK. Plant–microbe competition for soil amino acids in the alpine tundra: effects of freeze–thaw and dry–rewet events. Oecologia. 1998;113:406–14.

Lipson DA, Schmidt SK, Monson RK. Carbon availability and temperature control the post–snowmelt decline in alpine soil microbial biomass. Soil Biol Biochem. 2000;32:441–8.

Liu J, Sun X, Yu J. Nitrogen content variation in litters of *Deyeuxia angustifolia* and *Carex lasiocarpa* in Sanjingplai. Chin J Appl Ecol. 2000;11(6):898–902 (in Chinese).

Liu J, Wang J, Li Z, et al. N2O concentration and its emission characteristics in Sanjiang Plain wetland. Environ Sci. 2003;24(1):33–9 (in Chinese).

Logsdail DE, Webber LR. Effect of frost action on structure of Haldimand clay. Can J Soil Sci. 1959;39:103–6.

Ludwig B, Wolf I, Teepe R. Contribution of nitrification and denitrification to the emission of N2O in a freeze–thaw event in an agricultural soil. J Plant Nutr Soil Sci. 2004;167:678–84.

Lumbanraja J, Evangelou VP. Adsorption–desorption of potassium and ammonium at low cation concentrations in three Kentucky subsoils. Soil Sci. 1994;157:269–78.

Lv X. A review and prospect for wetland Science. Bull Chin Acad Sci. 2002;17(3):170–2 (in Chinese).

Lv X, Huang X. Advances in Wetland Study of China. Scientia Geographica Sinica. 1998;18(4):293–7 (in Chinese).

Ma X, Niu H. Mires in China. Beijing: Science Press; 1991 (in Chinese).

Macleod RA, Caleott RH. Cold shock and freezing damage to microbes. In: Gray TRG, Postage JR, editors. The survival of vegetative microbes. Cambrige: Cambridge University Press; 1976.

Magill AH, Aber JD. Variation in soil net mineralization rates with dissolved organic carbon addition. Soil Biol Biochem. 2000;32:597–601.

Matschonat G, Matzner E. Soil chemical properties affecting NH^{4+} sorption in forest soils. J Plant Nutr Soil Sci. 1996;159:505–11.

Mikan CJ, Schimel JP, Doyle AP. Temperature controls of microbial respiration in arctic tundra soils above and below freezing. Soil Biol Biochem. 2002;34:1785–95.

Moore TR, Souza WD, Koprivnjak JF. Controls on the sorption of dissolved organic carbon by soils. Soil Sci. 1992;2:120–9.

Ostroumov VE, Siegert C. Exobiological aspect of mass transfer in microzones of permafrost deposits. Adv Space Res. 1996;18:79–86.

Othman MA, Benson CH. Effect of freeze–thaw on the hydraulic conductivity and morphology of compacted clay. Can Geotech J. 1993;3:236–46.

Oztas T, Fayetorbay F. Effect of freezing and thawing processes on soil aggregate stability. Catena. 2003;52:1–8.

Panoff JM, Thammavongs B, Gueguen M. Cold stress responses in mesophilie bacteria. Cryobiology. 1998;36:75–83.

Peltovuori T, Soinne H. Phosphorus solubility and sorption in frozen, air-dried and field-moist soil. Eur J Soil Sci. 2005;56:821–6.

Phillips IR. Nitrogen availability and sorption under alternating waterlogged and drying conditions. Commun Soil Sci Plant Anal. 1999;30:1–20.

Prieme A, Christensen S. Natural perturbations, drying–rewetting and freeze–thawing cycles, and the emission of nitrous oxide, carbon dioxide and methane from farmed organic soils. Soil Biol Biochem. 2001;33:2083–91.

Qu X, Jia H, Zhang H, Li P. Preliminary study on purification function of reed wetland for nutrients from land sources. Chin J Appl Ecol. 2000;11(2):270–2 (in Chinese).

Raulund-Rasmussen K, Borrggaard OK, Hansen HCB, Olsson M. Effect of natural soil solutes on weathering rates of soil minerals. Eur J Soil Sci. 1998;49:397–406.

Rover MO, Heinemeyer KEA. Microbial induced nitrous oxide emissions from an arable soil during winter. Soil Biol Biochem. 1998;30:1859–65.

Schimel JP, Clein JS. Microbial response to freeze–thaw cycles in tundra and taiga soils. Soil Biol Biochem. 1996;28:1061–6.

Shen YH. Sorption of natural dissolved organic matter on soil. Chemosphere. 1999;38:1505–15.

Six J, Bossuyt H, Degryse S, et al. A history of research on the link between (micro) aggregates, soil biota, and soil organic matter dynamics. Soil Tillage Res. 2004;79:7–31.

Skogland T, Lomeland S, Goksoyr J. Respiratory burst after freezing and thawing of soil: experiments with soil bacteria. Soil Biol Biochem. 1988;20:851–6.

Soinne H, Peltovuori T. Extractability of slurry and fertilizer phosphorus in soil after repeated freezing. Agric Food Sci. 2005;14:181–8.

Sommerfeld RA, Mosier AR, Musselman RC. CO2, CH4 and N2O flux through a Wyoming Snow Pack and implications for global budgets. Nature. 1993;361:141–2.

Song C, Yan B, Wang Y, et al. CO2 and CH4 Fluxes and their influencing factors in Sanjiang Plain wetlands. Chin Sci Bull. 2003;48(23):2473–7 (in Chinese).

Song C, Wang Y, Yan B, et al. The changes of the soil hydrothermal condition and the dynamics of C, N after the mire tillage. Environ Sci. 2004;25(3):150–4 (in Chinese).

Song C, Wang Y, Wang Y, et al. Dynamics of CO2, CH4 and N2O emission fluxes from mires during freezing and thawing season. Environ Sci. 2005a;26(4):7–12 (in Chinese).

Song C, Wang Y, Wang Y, et al. Difference of soil respiration and CH4 flux between mire and areble soil. Chin J Soil Sci. 2005b;36(1):45–9 (in Chinese).

Soulides DA, Allison FE. Effect of drying and freezing soils on carbon dioxide Production. Soil Sci. 1961;91:291–8.

Sulkava P, Huhta V. Effects of hard frost and freeze–thaw cycles on decomposer communities and N mineralisation in boreal forest soil. Appl Soil Ecol. 2003;22:225–39.

Sun Z. Study on the nitrogen biogeochemical process of *Calamagrostis angustifolia* Wetland in the Sanjiang Plain. Graduate University of Chinese Academy of Sciences. 2007 (in Chinese).

Teepe R, Brumme R, Beese F. Nitrous oxide emissions from soil during freezing and thawing periods. Soil Biol Biochem. 2001;33:1269–75.

Thompson TL, Blackmer AM. Quantity–Intensity relationships of soil ammonia in long-term rotation plots. Soil Sci Soc Am J. 1992;56:494–8.

Tong C, Zhang W, Wang H. Relationship between organic carbon and water content in four type wetland sediments in Sanjiang Plain. Environ Sci. 2005;26(6):38–42.

Ussiri DAN, Johnson CE. Sorption of organic carbon fractions by Spodosol mineral horizons. Soil Sci Soc Am J. 2004;68:253–62.

Vandenbruwane J, De Neve S, Qualls RG, Sleutel S, Hofman G. Comparison of different isotherm models for dissolved organic carbon (DOC) and nitrogen (DON) sorption to mineral soil. Geoderma. 2007;139:144–53.

Vaz MDR, Edwards AC, Shand CA, Cresser MS. Changes in the chemistry of soil solution and acetic-acid extractable-P following different types of freeze–thaw episodes. Eur J Soil Sci. 1994;45:353–9.

Viklander P. Permeability and volume changes in till due to cyclic freeze–thaw. Can Geotech J. 1998;35(3):471–7.

Walker VK, Palmer GR, Voordouw G. Freeze–thaw tolerance and clues to the winter survival of a soil community. Appl Environ Microbiol. 2006;72:1784–92.

Wang FL, Alva AK. Ammonium adsorption and desorption in sandy soils. Soil Sci Soc Am J. 2000;64:1669–74.

Wang FL, Bettany JR. Organic and inorganic nitrogen leaching from incubated soils subjected to freeze–thaw and flooding. Can J Soil Sci. 1994;74:201–6.

Wang C. Experimental study of nitrogen transport and transformation in soils. Adv Water Sci. 1997;8(2):176–82 (in Chinese).

Wang J, Xu X, Wang Y. Thermal sieve effect and convectional migration of soil particles during unidirectional freezing. J Glaciol Geocryol. 1996;18(3):252–5 (in Chinese).

Wang R, Zhang S, Qin G. Effects on freezing–tahwing properties of fresh concrete by fine ground slag. J Huainan Inst Technol. 2001;21(4):35–8 (in Chinese).

Wang D, Lv X, Ding W, et al. Comparison of methane emission from marsh and paddy field in Sanjiang Plain. Scientia Geographica Sinica. 2002a;22(4):500–3 (in Chinese).

Wang D, Lv X, Ding W, et al. Methane emission from marshes in Zoige Plateau. Adv Earth Sci. 2002b;17(6):877–80 (in Chinese).

Wang G, Liu J, Wang J, Yu J. Soil phosphorus forms and their variations in depressional and riparian freshwater wetlands (Sanjiang Plain, Northeast China). Geoderma. 2006;132:59–74.

Wang G, Liu J, Zhao H, Wang J, Yu J. Phosphorus sorption by freeze-thaw treated wetland soils derived from a winter-cold zone. Geoderma. 2007;138:153–61.

Williams CF, Agassi M, Letey J. Facilitated transport of napropamide by dissolve organic matter through soil columns. Soil Sci Soc Am J. 2000;64:590–4.

Winter JP, Zhang ZY, Tenuta M, Voroney RP. Measurement of microbial biomass by fumigation–extraction in soil stored frozen. Soil Sci Soc Am J. 1994;58:1645–51.

Xin XL, Xu FA, Zhang JB, Xu MX, Zhu AN. A new instrument for monitoring soil moisture in field. Soils. 2005;37:326–9 (in Chinese).

Yanai Y, Toyota K, Okazaki M. Effects of successive soil freeze–thaw cycles on soil microbial biomass and organic matter decomposition potential of soils. Soil Sci Plant Nutr. 2004;50:821–9.

Yang P, Zhang T. The physical and the mechanical properties of original and frozen–thawed soil. J Glaciol Geocryol. 2002;24(5):665–7 (in Chinese).

Yang J, Yu J, Liu J, et al. N2O and CH4 fluxes in an island forest in wetland, Sanjiang Plain. Ecol Environ. 2004;13(4):476–9 (in Chinese).

Yin C. The ecological function, protection and utilization of land/inland water ecotones. Acta Ecologia Sinica. 1995;15(3):331–5 (in Chinese).

Yu XF, Zhang YX, Zhao HM, et al. Freeze–thaw effects on sorption/desorption of dissolved organic carbon in wetland soils. Chin Geogr Sci. 2010;20(3):209–17.

Yu XF, ZhangY X, Zou YC, et al. Sorption and desorption of ammonium in wetland soils subject to freeze-thaw. Pedosphere. 2011a;21(2):251–8.

Yu XF, Zou YC, Jiang M, et al. Response of soil constituents to freeze-thaw cycles in wetland soil solution. Soil Biol Biochem. 2011b;43(6):1308–20.

Zhu Z, Wen X. Soil nitrogen in China. Nanjing: Jiangsu Science & Technology Press; 1992 (in Chinese).

Chapter 2
Overview of the Study Area

2.1 Natural Conditions of the Study Area

2.1.1 Location

The Sanjiang Plain is located in the northeast corner of Heilongjiang Province, northeast China. It is a marshy low plain mainly formed by the alluvial confluence of the Heilongjiang, Songhua and Ussuri rivers, and is one of China's largest contiguous distribution areas of freshwater wetlands. It is also greatly affected by human activities, and is a region with one of the fastest reductions in natural wetland area (Liu and Ma 2002).

The Sanjiang Plain includes the low plains formed by alluvium from the Songhua, Heilong and Ussuri rivers to the north of Wanda Mountain and the plains formed by alluvium from the Ussuri River including its tributaries and Xingkai Lake to the south of Wanda Mountain. The northernmost latitude is 48°28′ and the southernmost latitude is 45°01′; the western edge is east longitude 130°13′ and the eastern edge is longitude 135°05′. The entire region includes 23 cities, counties, and 52 state farms and 8 forest industry bureaus, which are located in these cities and counties. The total area of the region is 1.089×10^5 hm^2, with a total area of plains of 51,300 hm^2 (not including the Wanda Mountain area). The setup of the sampling sites in this study and their locations on the Sanjiang Plain are shown in Fig. 2.1.

2.1.2 Topographical Features

The Sanjiang Plain is located in a Cenozoic inland rift basin, which later accumulated deep unconsolidated sediments from the Tertiary and Quaternary periods. Since the Quaternary period, neotectonic movement has resulted in intermittent

X. Yu, *Material Cycling of Wetland Soils Driven by Freeze–Thaw Effects*,
Springer Theses, DOI: 10.1007/978-3-642-34465-7_2,
© Springer-Verlag Berlin Heidelberg 2013

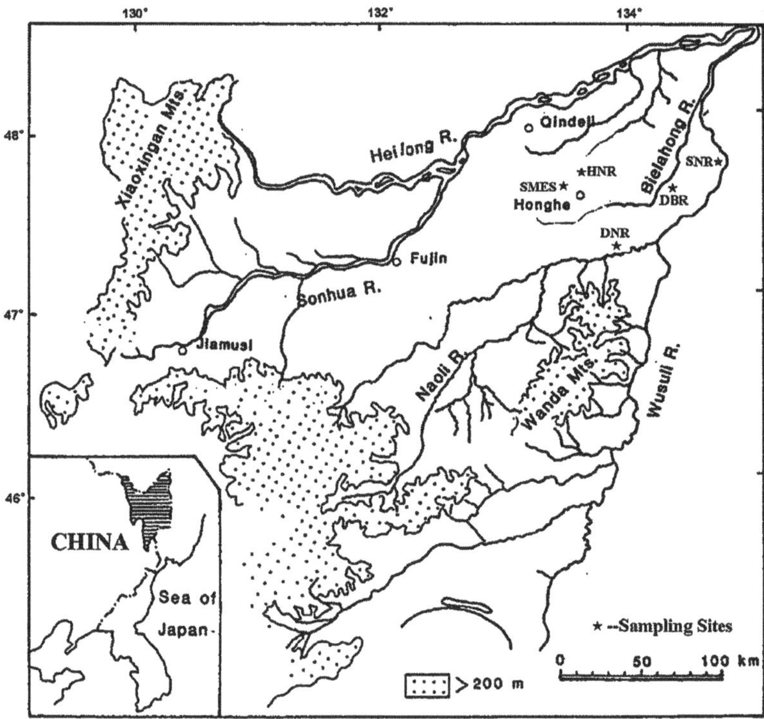

Fig. 2.1 Distribution of the study sites. *HNR* Honghe Nature Reserve, *SNR* Sanjiang Nature Reserve, *SMES* Sanjiang Mire Experimental Station, *DBR* downstream of Bielahong River, *DNR* Haiwang Village, downstream of Naoli River (modified after Wang et al. 2006, Fig. 1, and any original (first) copyright notice displayed with material)

and slow sinking occurring over a large area and unequal rising occurring only on the periphery and in some local regions. Because of unequal long-term sinking, together with long-term changing and flooding of the paths of the Heilongjiang, Songhua, and Ussuri rivers, meanders are well developed, river coast/banks and low-level terrain are very broad and significant development of marshes has occurred. Many marshes, abandoned river paths and oxbow lakes of different size and depth remain on the floodplain and terraces, especially sunk marshes from the melting of frozen soil, as the so-called hot melted lakes or melt sunk marshes. In addition, the watershed line between the tributaries is not obvious, and some rivers have no obvious riverbed or riverbeds are above the ground and therefore become aboveground rivers. The converged flooded water and surface runoff water have difficulty infiltrating the soil because of a sub-surface clay layer and sub-clay layer. Therefore surface drainage is very difficult, which causes most of the marshes to be covered with water all year and the long-term gleying of soils (Sun 1990; Zhao 1999).

2.1.3 Climate Conditions

This temperate zone area has a humid or sub-humid continental monsoon climate, with dry and cold winters, warm and rainy summers, fast-warming and windy springs, and sharply cooling autumns. According to statistical analysis, the annual average temperature is 1.9 °C, the annual effective accumulated temperature greater than or equal to 10 °C is 2300 °C, the annual rainfall is 500–650 mm, with 70 % concentrated from June to September, and the frost-free period is 120–125 days. Temperatures tend to be higher in the south and lower in the north, while precipitation tends to be higher in the east and less in the west. Evaporation is low in the rainy summer and the land surface may evaporate 550–650 mm annually. The potential evaporation of this area is less than precipitation. Therefore, the moist index in this area is low, being less than 1 in the eastern region and 1.0–1.1 in the western region. More precipitation and less evaporation provides the water source for the formation of marshes in this area, which has a long frozen period with the frozen layer being completely melted at the end of June. Because of the low thermal conductivity of the grassroots layer or the peat layer, wetland areas, especially peat marsh areas, are often not completely melted until July or early August, which promotes the development of wetlands (Cai 1990; Chen 1996).

2.1.4 Hydrological Characteristics

The Sanjiang Plain has sparse rivers and a low river network density. Except for the three major rivers, most rivers are small and medium-sized. The main rivers in this area are the Songhua, Heilongjiang and Ussuri rivers and their tributaries (Bielahong River, Naoli River etc.). Major lakes are the Great Xingkai Lake and the Mini Xingkai Lake. A large area of stagnant water has been formed and marsh development promoted because of overflowing of broad floods caused by backwater lift and water terrace detention on the surface. Wetland development has been enhanced by the poor drainage (Cai 1990; Chen 1996).

2.1.5 Distribution of Vegetation

Plant species of the Sanjiang Plain belong to the flora of Changbai Mountain. The watery and over-wet environment provides favorable conditions for the growth and reproduction of marsh and wetland plants. Roots of densely clustered marsh or wetland plants are intertwined and generally form a 20–30 cm thick root layer, with a high water-holding capacity. Ecological and geographical conditions that constrain vegetation formation are the low and flat terrain, the heavy and sticky clay soil, the high water content, the cold and long winters, and the warm and short summers. Thus, the constructive plants, the dominant plants and the main associated plants in this area, are all water-adaptive helophytes, hygrophytes and a few mesophytes, mainly including *Carex lasiocarpa, C. meyeriana, C. pseudocuraica, Calamagrostis angustifolia, Betula fruticosa,* and *Salix rosmarinifolia var. brachypoda,* in association with *Iris laevigata, Menyanthes trifoliata, Caltha palustris, Lythrum salicaria, Pedicularis grandiflora, Verairum dahuricum, Sanguisorba parviflora* and other species. Meadow vegetation is distributed on the high floodplain, terraces and piedmont sloping plain, mainly including *Calamagrostis angustifolia* meadow and *Betula fruticosa*-miscellaneous meadow (Zhao 1999).

2.2 Wetland Soil Background of the Studied Area

2.2.1 Characteristics of Wetland Soil Distribution

The main wetland soils in the Sanjiang Plain include three different types of wetland: albic soils, marsh soils and peat soils. The largest area of soils on the Sanjiang Plain is albic soils which are mainly distributed in areas of relatively high terrain. Marsh and peat soils are mainly distributed in the poorly drained marsh environment, such as river, low flood land, high flood land and the loop-type swale on the low-level ground. The soils can be further divided into more than one soil class. For example, from high to low topography, the albic soils can be divided into brown white albic soils, meadow albic soils and gley albic soils, the marsh soils includes peat marsh soils, meadow marsh soils, humus marsh soils and sapropel marsh soils, and the peat soils include thick layer meadow peat soils, middle layer meadow peat soils and thin layer meadow peat soils. Many of these soils have been affected by human activities and reclaimed into farmland and thus have become artificial wetland soils (Liu et al. 2004).

2.2.2 Physical Characteristics of the Soils

Wetland soils on the Sanjiang Plain are mainly low-salinity freshwater wetland soils, and most of them have high organic matter content. The soils are loose with many pores, a high moisture content and low unit surface density, with bulk density values of 0.1–0.8 g cm^{-3} and specific density of 1.60–2.40 g cm^{-3} with most of the specific gravity values of organic soils being below 2.0 g cm^{-3}, while values for mineral wetland soils are often greater than 2.0 g cm^{-3}. Variation in unit densities of Sanjiang Plain wetland soils occurs because of the very different organic matter contents in the different types of wetland soil. For peat soils and peat marsh soils, plant residue composition in the peat layer is related to the degree of decomposition of the peat. The unit weights of peat layers composed of different plant residues are significantly different, with the unit weight of residues from mosses and herbs being the lowest. The unit weight of *Calamagrostis angustifolia* communities is the highest at 0.8 g cm^{-3}, followed by the *Carex lasiocarpa-Glyceria angustifolia* plant communities.

The wetland soils on the Sanjiang Plain have relatively heavy textures and are sticky. From the analysis of mechanical composition, in white slurry soil more than 55 % of the physically sticky particles are less than 0.01 mm in size, and 25–40 % are less than 0.005 mm in size. The proportions of particles and physically sticky particles decrease with increasing soil depth. In marsh surface soils about 20 % of the physically sticky particles are less than 0.01 mm in size, with the percentage increasing with depth to more than 25 % at a depth of 80 cm. In the surface layer about 10 % of the sticky particles are less than 0.005 mm in size, increasing to 70 % at a depth of 80 cm (Yang et al. 2004).

2.2.3 Chemical Properties of the Soils

Organic matter in wetland soils includes the grass-roots layer of the soil, the humus layer, and the peat layer. Plant residues in soil are continuously decomposed and transformed by microorganisms, finally forming humus or peat layers. The process of formation of humus or peat is a major component of the formation of wetland soils, and it is also closely related to soil development and soil fertility capacity. The organic content in the surface layer of Sanjiang Plain soils is high, with most of them exceeding 10 % although the soil organic content varies between different types of wetlands. The surface organic content of the surface soil layer is between 8 and 10 % for meadow albic soils, about 10 % for gley albic soils, 15–20 % for humus marsh soils, and up to 30 % for muddy marsh soils.

The exchange capacity and pH of soils results from the interaction of comprehensive factors during the soil formation process, and are important indicators of soil fertility. The exchange capacity of the soil surface layer of the Sanjiang Plain wetlands is larger, with a cation-exchange capacity of 20 me/100 g of soil and a base-exchange capacity of 20–40 me/100 g of soil. The exchange capacity sharply reduces below the surface, with the cation-exchange capacity being 10–20 me/100 g of soil in the transition layer. For the parent material layer, the cation-exchange capacity increased to about 20 me/100 g of soil with a base-exchange capacity of 25 me/100 g of soil. Nitrogen, phosphorus and potassium are indispensable nutrient elements for plant growth and are also components of all proteins and protoplasm. Therefore, the amounts of these elements will have direct impacts on plant production yields. Most nitrogen concentrations in wetland soils of the Sanjiang Plain are above 0.5 %, while the total phosphorus concentration is low, and the total potassium concentrations do not vary significantly. Nitrogen, phosphorus and potassium concentrations in wetland soils are relatively high, but most of them are in organic form, which cannot be directly absorbed and used by crops. The soils have to be ripened so that the organic form of the elements can be released to the soil through mineralization and be utilized by crops, which indicates that the potential fertility of the wetland soils is relatively high (Yang et al. 2004).

2.3 Chapter Summary

In this chapter, the location, topography, hydrologic conditions, climate and vegetation characteristics of the study area, the Sanjiang Plain, are summarized. As China's largest contiguous distribution area of freshwater wetlands, the Sanjiang Plain is an ideal platform to study freeze–thaw processes in marsh wetlands. Through literature and field monitoring, soil environmental characteristics of marsh wetland in the study area (including soil distribution and soil physical, chemical and biological characteristics) were discussed. This provides background information on the regional ecological environment for the subsequent experimental study.

References

Cai X. Pealand science. Beijing: Geology Press; 1990 (in Chinese).

Chen G. Study on marsh in the Sanjiang Plain. Beijing: Science Press; 1996 (in Chinese).

Liu H, Lv X, Zhang S. Landscape biodiversity of wetlands and their changes in 50 years in watersheds of the Sanjiang Plain. Acta Ecologica Sinica. 2004;24(7):1472–9 (in Chinese).

Liu X, Ma X. Natural environmental change and ecological conservation in Sanjiang Plain. Beijing: Science Press; 2002 (in Chinese).

Sun G. Preliminary study on formation and evolution of geomorphology of the Sanjiang Plain in Sanjiang Plain. Harbin: Harbin Map Publishing House; 1990 (in Chinese).

Wang G, Liu J, Wang J, Yu J. Soil phosphorus forms and their variations in depressional and riparian freshwater wetlands (Sanjiang Plain, Northeast China). Geoderma. 2006;132:59–74.

Yang Q, Liu J, Lv X. Structure and function of soil vegetation–animal system of annular wetland in the Sanjiang Plain. Chin J Ecolo. 2004;23(4):72–77 in Chinese.

Zhao K. Mires of China. Beijing: Science Press; 1999 (in Chinese).

Chapter 3
Freeze–Thaw Principles of Wetland Soils

The freeze–thaw process is an important characteristic of climate change in seasonal freeze–thaw wetlands, which affects the biogeochemical processes of wetland ecosystems in many ways (Piao and Liu 1995). As a type of abiotic stress, freeze–thaw can change the soil aggregates (Oztas and Fayetorbay 2003), permeability (Chamberlain and Gow 1979; Kim and Daniel 1992) and mechanical properties (Wang et al. 2001). The freeze–thaw process also indirectly affects nutrient circulation, through its impact on soil microbial activity (Bolter 2005), gas release (Chang and Hao 2001) and the concentrations of dissolved carbon and nitrogen in soil (Wang and Bettany 1994; Grogan et al. 2004; Wang et al. 2007).

The Sanjiang Plain is a seasonal freeze–thaw region which has a long frozen period. Temperatures decline sharply in the autumn, and it begins to freeze in late October and to melt in early April of the following year. Thus the frozen period lasts for up to 6 months and the average freezing depth is 150–210 cm (Ma and Niu 1991). The freeze–thaw process in this region not only varies with season but also varies daily, and the complex, long-term and significant freeze–thaw effects will inevitably affect the biogeochemical processes in wetlands.

Using field monitoring over two consecutive years, temperature changes in different soil layers, soil overlying water and snow were analyzed in this chapter, and the temperature of the experimental zone was increased by using a film greenhouse in order to simulate the soil freeze–thaw situation that might arise because of global warming.

3.1 The Variation of Wetland Ice and Snow During the Freeze–Thaw Period

The field experiment was performed at the Sanjiang Mire Wetland Experimental Station, Chinese Academy of Sciences (47°35′N, 133°31′E), in the Sanjiang Plain.

X. Yu, *Material Cycling of Wetland Soils Driven by Freeze–Thaw Effects*,
Springer Theses, DOI: 10.1007/978-3-642-34465-7_3,
© Springer-Verlag Berlin Heidelberg 2013

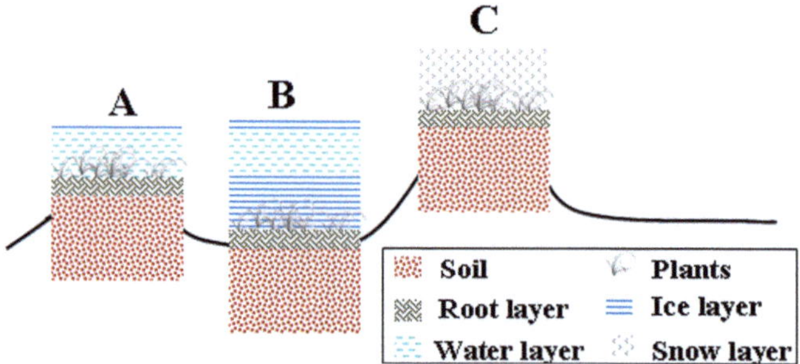

Fig. 3.1 Characteristics of freeze and thaw in Sanjiang Plain (with kind permission from IEEE: Yu and Wang 2008, Fig. 1, and any original (first) copyright notice displayed with material)

For the artificial simulation of global warming, the studying areas were selected in the *Calamagrostis angustifolia* dominated marsh which has a varied micro-topography (Fig. 3.1). The ice observation plot (1 × 1.2 m, film covered, short for FC), was set in the seasonal waterlogged position (B). The thin snow on the ice was cleared and then covered with the plastic transparent film with a thickness of 0.01 mm. The adjacent plot without film was observed as a control to the experiment plot (uncovered, short for UC). The thickness of ice was measured from the date after plot design (5 April, 2007) to the date when the ice completely thawed (9 April, 2007). The snow observation plot in the seasonal dry position (C) was designed in the same way and the surface snow temperature was measured during the same periods. Each plot had two replicates.

The instrument used to detect the temperature was Tri-Sense Kit Model: 37000-95 (Manufacturer: Cole-Parmer, USA), which provides highly accurate (± 0.1 % of reading) measurements. The ice thickness and soil depth were measured with a steel ruler.

The FC ice thawed more quickly than the UC ice (Fig. 3.2). With one-way ANOVA analysis based on the mean thickness during the thawing periods, the differences between plots 1, 2, and 3 were not significant for FC and UC ($F = 0.147$, $p = 0.865$; $F = 0.231$, $p = 0.798$, respectively). According to the independent t test, the means of all ice thickness regardless of plot difference between FC and UC had a significant difference ($t = -2.359$, $p = 0.028$) (Fig. 3.3). Compared with UC, the ice thickness of FC decreased by 35.29 %. Furthermore, the relative ice thaw rate, calculated from the daily difference of ice thickness ($(t_{i+1} - t_i)/t_i$), showed that the rate of FC decreased gradually while that of UC fluctuated. This result indicated that the film may mitigate the ice thaw process significantly.

The temperature of the surface snow of FC was higher than that of UC. It increased quickly during the first 2 days, after which it increased at a lower rate (Fig. 3.4). According to the independent t test, the difference of surface snow temperature between FC and UC was not significant ($t = 0.836$, $p = 0.408$), and the means were 0.14 ± 0.25 °C, -0.26 ± 0.21 °C, respectively.

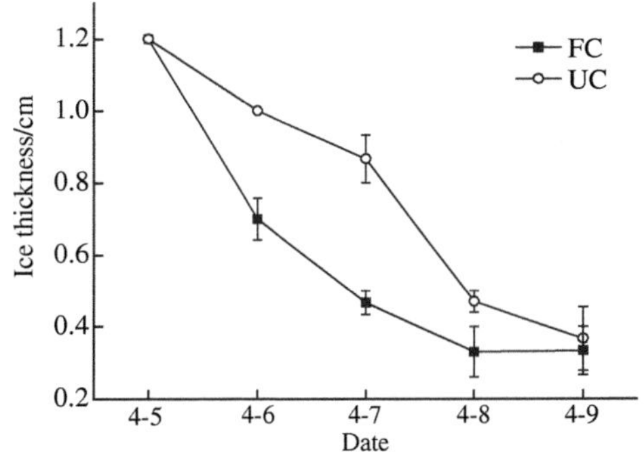

Fig. 3.2 Variation of ice thickness (with kind permission from IEEE: Yu and Wang 2008, Fig. 2, and any original (first) copyright notice displayed with material)

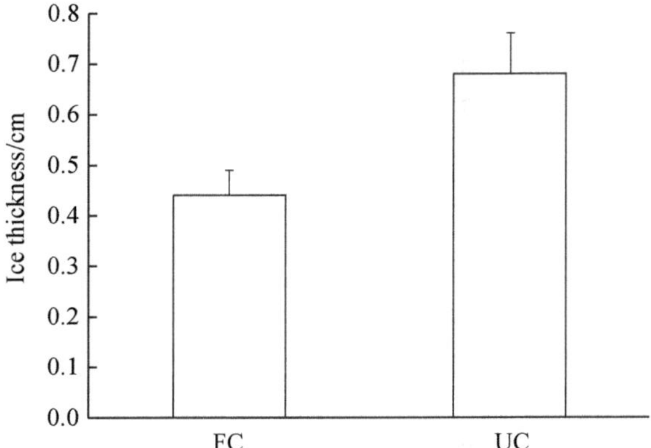

Fig. 3.3 Means of ice thickness for FC and UC (M ± S.E.) (with kind permission from IEEE: Yu and Wang 2008, Fig. 3, and any original (first) copyright notice displayed with material)

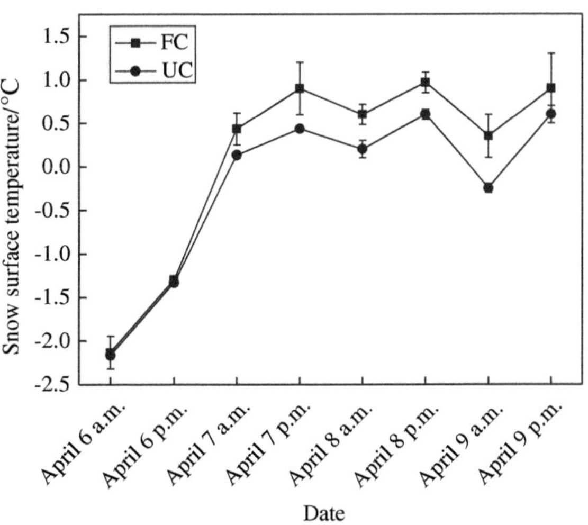

Fig. 3.4 Variation of snow surface temperature (with kind permission from IEEE: Yu and Wang 2008, Fig. 4, and any original (first) copyright notice displayed with material)

3.2 Wetland Soil Temperature Variation During the Freeze–Thaw Period

3.2.1 Wetland Soil Temperature Changes with Water Depth and Thaw Depth

The thaw characteristic of the soil beneath the melt water was observed. Table 3.1 shows that when the water was deeper, the water temperature was lower, and the thaw depth of soil was deeper.

3.2.2 The Relationship Between Soil Thaw Depth and Soil Temperature

Three identical plots (1 × 1.2 m) in drained agriculture field soil were selected and the soil profile temperature was measured, as well.

At 19 April 2007, the thaw depth of reclaimed soils was 30–40 cm, while the thaw depth of the marsh soils was about 15–20 cm. The study's results show that at farmland, the temperature of a certain thawed soil layer was related to the distance from that layer to the frozen borderline. The relationship between the distances and the average temperatures of three thawed soil layer replicates was described by the logistic curve illustrated in Fig. 3.5. According to the function, the temperature of any soil layer in this area can be estimated.

The water depth would be deeper because of the accelerated thaw process of ice and snow. Therefore, it can influence the freeze–thaw process of the soils beneath the thawed water, which has been mentioned earlier. More and more marsh soils have been reclaimed for farm soils. This seriously disturbed the balance of the freshwater marsh ecosystem. The change of the soil freeze–thaw process is one aspect. When the marsh soils were reclaimed for farmland soils, many kinds of plants were replaced by crops, and the water was drained. The soils without the mitigation of the water and plants must freeze and thaw quicker than marshes soils.

Table 3.1 Variations of water depths and temperatures of water and thawed soil

Date	Water depth (cm)			Water temperature (°C)			Temperature of thawed soil (°C)		
	1#	2#	3#	1#	2#	3#	1#	2#	3#
14 April	20	15	5	10.3	12.4	14.7	3	5	10
15 April	20	15	5	12.5	14.2	16.6	5	8	13
16 April	19	12	3	15.1	16.3	18.3	10	10	20
17 April	18	11	2	13.2	14.6	17.3	12	13	24
18 April	18	9	2	10.4	11.6	16.4	12	14	25
19 April	15	8	2	14.2	15.2	15.3	15	20	25

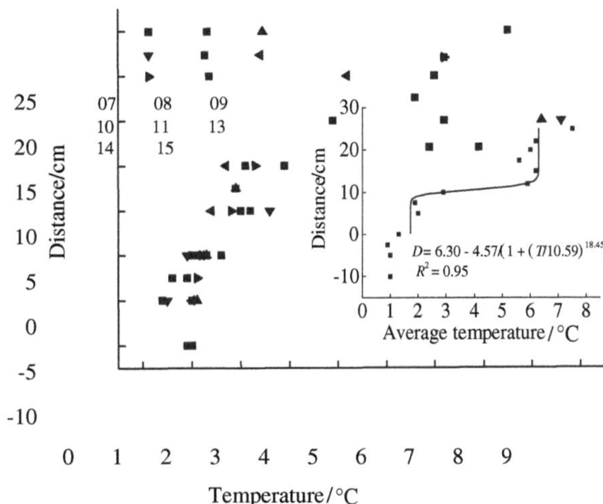

Fig. 3.5 Relationship of soil temperature and the distance between the soil layer and the freeze borderline (with kind permission from IEEE: Yu and Wang 2008, Fig. 5, and any original (first) copyright notice displayed with material)

3.2.3 Temperature Variation in Wetland Soil Layers During the Freeze–Thaw Period

The field observation zone was located in a typical *Calamagrostis angustifolia* meadow in a marsh observation field at the Sanjiang Plain Marsh Wetland Ecological Experimental Station (47°35′N, 133°31′E). *Calamagrostis angustifolia* is the dominant species in this meadow, which is well drained except during the flooded season. Two comparison observation points, one with vegetation-covered soil and the other without vegetation-covered soil, were established in the observation zone. In order to simulate global warming, a third observation point was covered with seedling film greenhouses (2 × 1 m); to compare the effect of reclamation, two field observation points were set up in the middle of a dryland observation field at the Experimental Station.

Soil layer temperatures in the autumn freeze–thaw period were observed from November 5, 2007 to December 19, 2008, and soil layer temperatures in the spring freeze–thaw period were observed from March 13, 2008 to April 22, 2008. During these observation periods, two measurements were performed daily at 8:00 a.m. and 2:00 p.m. with a Tri-Sense Kit Model 37000-95 (Manufacturer: Cole-Parmer, USA). The accuracy is up to ±0.1 %.

Results of wetland soil layer temperatures in the 2007 autumn freeze–thaw period show that soil layer temperature decreased with fluctuations. The soybean growing farmland (reclaimed from wetlands) had the lowest soil temperature, followed by the wetland without vegetation cover, the wetland with vegetation cover and the wetland under the film greenhouse (Fig. 3.6). With the reclamation of wetland to farmland, the cooling trend of soil temperature in the autumn freeze–thaw period was severe. Soil temperature increased with increasing soil depth.

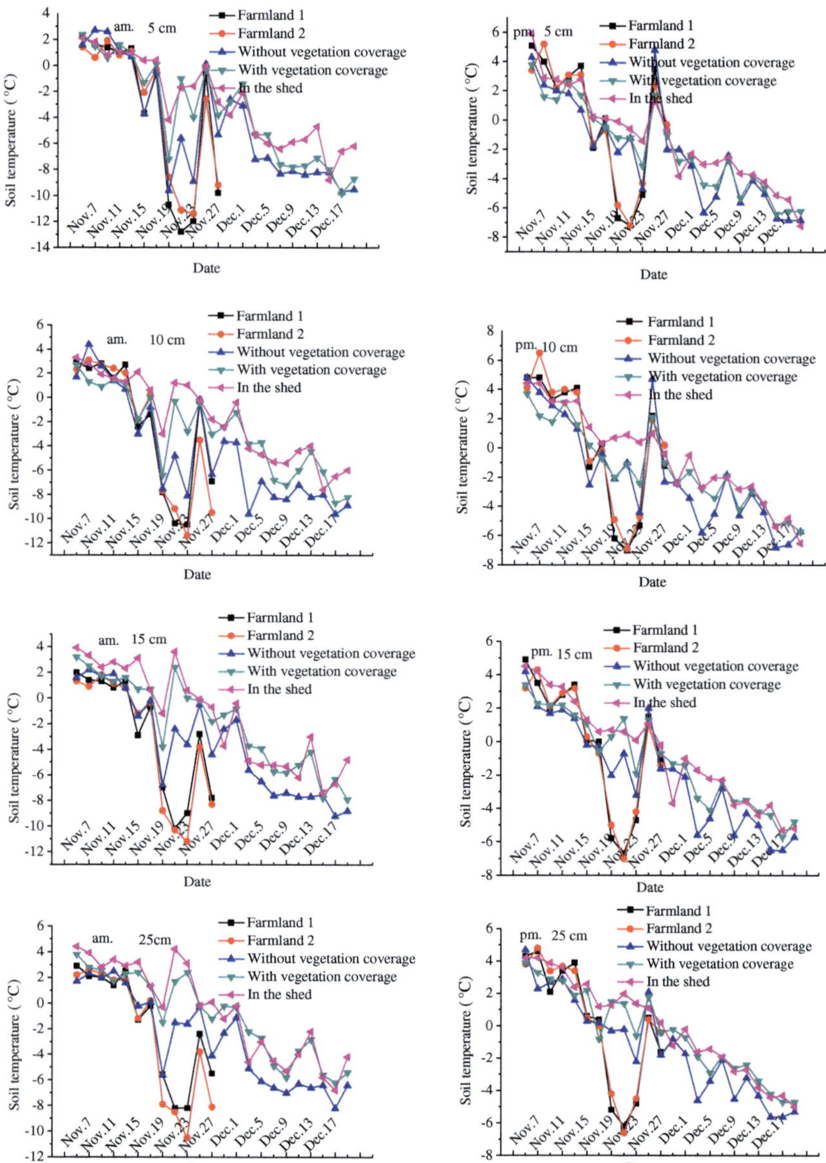

Fig. 3.6 Variation of soil temperature with time in autumn freeze–thaw season

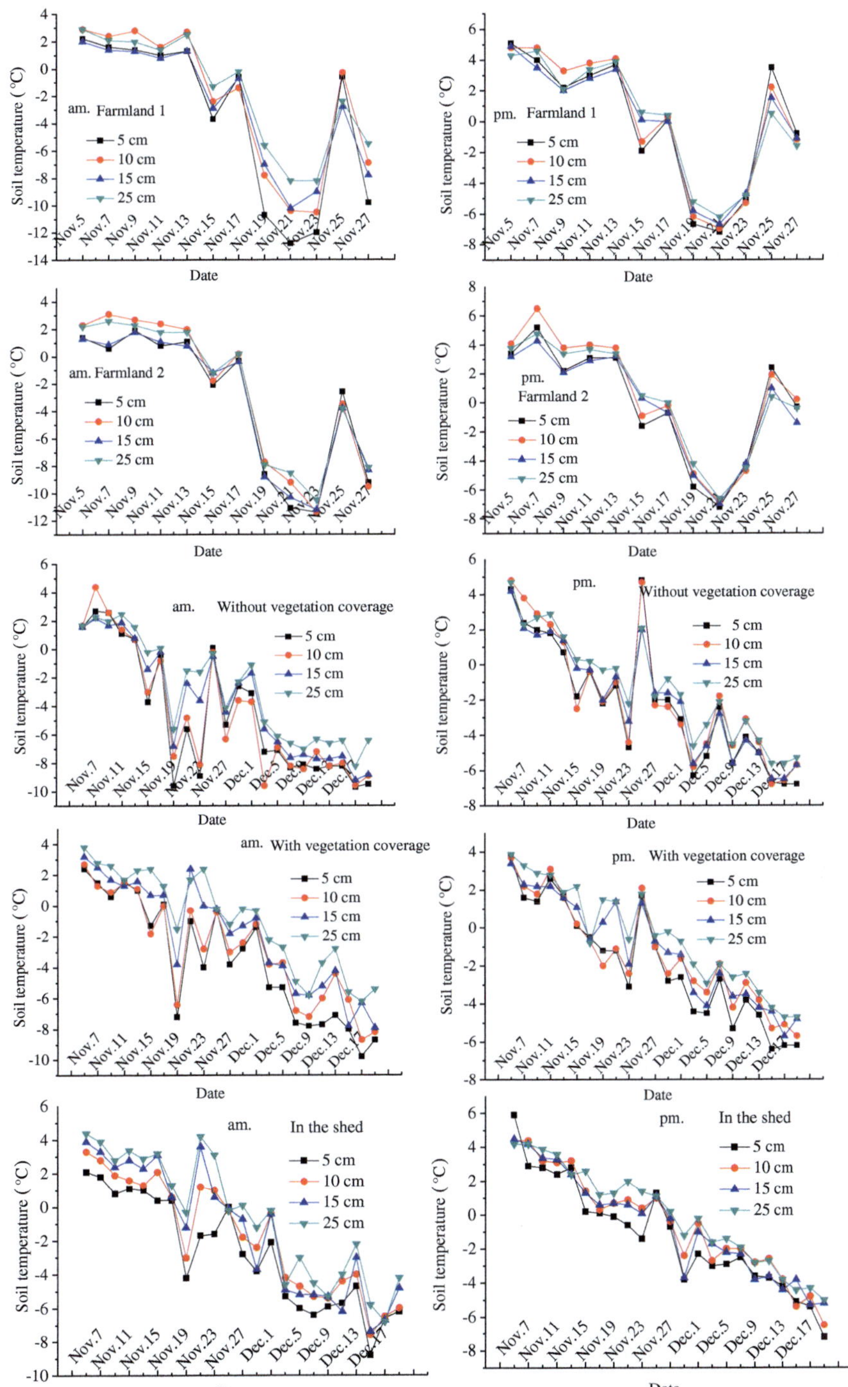

Fig. 3.7   Soil temperatures of different soil layers in autumn freeze–thaw season

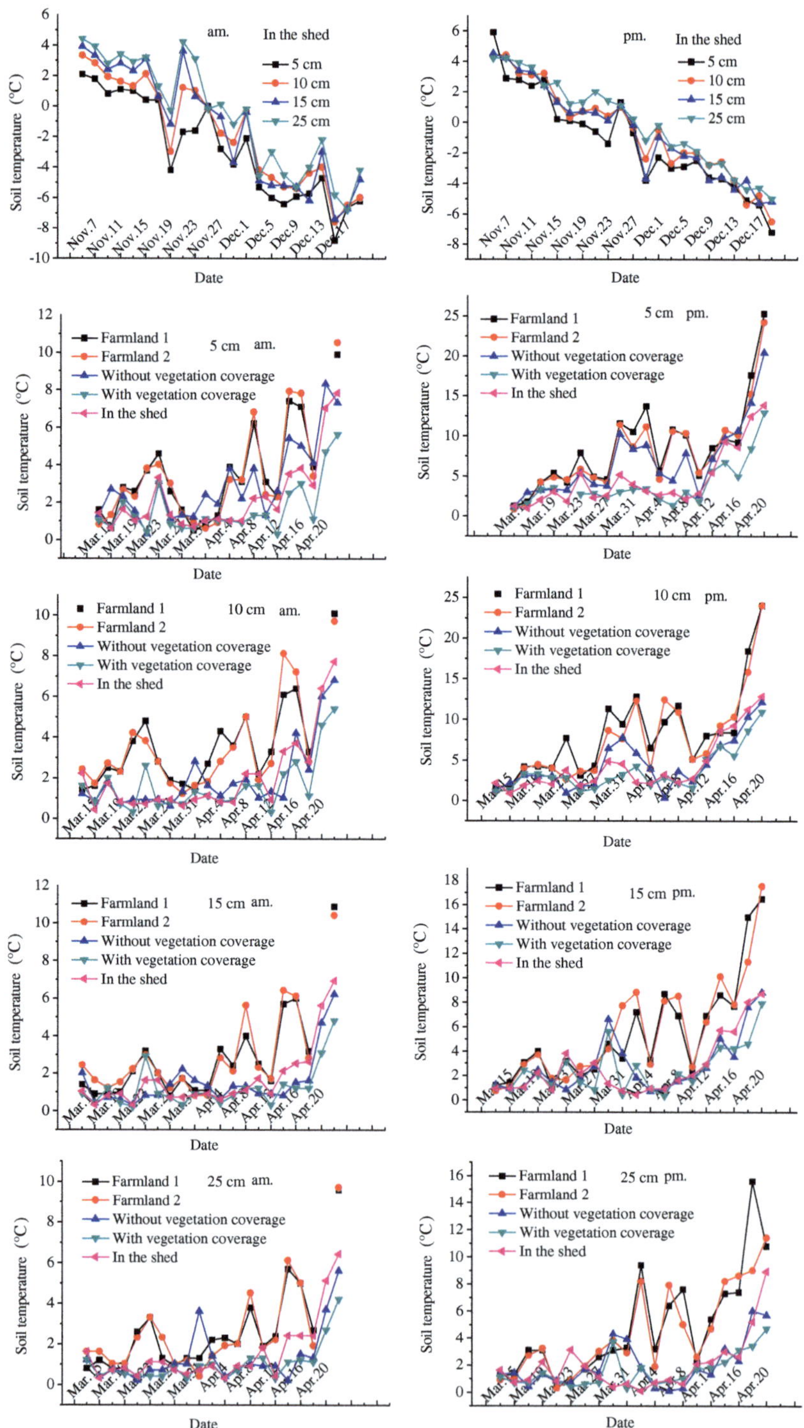

Fig. 3.8 Variation of soil temperature with time in spring thawing season

Figure 3.6 shows that, from November 13 to November 27, the daily temperature of the surface soil within 5 cm fluctuated around 0 °C, which indicates the presence of a daily freeze–thaw cycle. Comparison of soil temperature changes at different depths shows that soil temperatures in the vegetation-covered wetland and shed wetland increased with increasing soil depth (Fig. 3.7). For farmland and wetland without vegetation cover, the above-mentioned result was true only after they were frozen (temperature below 0 °C). Before the soil was frozen, soil temperature showed no obvious regularity.

Soil layer temperature observation results in the 2008 spring freeze–thaw period showed that the temperature of all soil layers had an upward trend with fluctuations. Farmland soil temperature was the highest, followed by wetland without vegetation cover, wetland under the film greenhouse and wetland with vegetation cover (Fig. 3.8). With the reclamation of wetland to farmland, the trend of increasing soil layer temperature in the spring freeze–thaw period was much more severe.

3.3 Chapter Summary

In this chapter, through continuous field monitoring and warming simulation, the freeze–thaw process of typical Sanjiang Plain wetland and farmland under the situation of current and simulated future climate warming was quantitatively studied. The presence of a film greenhouse simulating global warming directly accelerated the melt rate of frozen soil, and promoted the melt of soil-covered snow and ice, also indirectly affecting the soil melt process. With the same amount of snow precipitation, an accelerated snow melt rate will affect the accumulated water depth of marsh surface water, and will further affect the water temperature and freeze–thaw process of the underwater soil. With the reclamation of wetland to farmland, increases and decreases in soil temperature in farmland soils are more severe than in wetland.

References

Bolter M, Soethe N, Horn R, Uhlig C. Seasonal development of microbial activity in soils of northern Norway. Pedosphere. 2005;15:716–27.

Chamberlain EJ, Gow AJ. Effect of freezing and thawing on the permeability and structure of soils. Eng Geol. 1979;13:73–92.

Chang C, Hao X. Source of N_2O emission from a soil during freezing and thawing. Phyton Ann Rei Botanicae. 2001;41:49–60.

Grogan P, Michelsen A, Ambus P, Jonasson S. Freeze–thaw regime effects on carbon and nitrogen dynamics in sub-arctic heath tundra mesocosms. Soil Biol Biochem. 2004;36:641–54.

Kim WH, Daniel DE. Effects of freezing on hydraulic conductivity of compacted clay. J Geotech Eng. 1992;118:1083–97.

Ma X, Niu H. Mires in China. Beijing: Science Press; 1991 (in Chinese).

Oztas T, Fayetorbay F. Effect of freezing and thawing processes on soil aggregate stability. Catena. 2003;52:1–8.

Piao H, Liu G. Effects of alternative drying–rewetting and freezing–thawing on soil fertility and ecological environment. Chin J Ecol. 1995;14(6):29–34 (in Chinese).

Wang FL, Bettany JR. Organic and inorganic nitrogen leaching from incubated soils subjected to freeze–thaw and flooding. Can J Soil Sci. 1994;74:201–6.

Wang R, Zhang S, Qin G. Effects on freezing–tahwing properties of fresh concrete by fine ground slag. J Huainan Inst Technol. 2001;21(4):35–8 (in Chinese).

Wang G, Liu J, Zhao H, Wang J, Yu J. Phosphorus sorption by freeze-thaw treated wetland soils derived from a winter-cold zone. Geoderma. 2007;138:153–61.

Yu XF, Wang GP. Field simulation the effects of global warming on the freeze-thaw process of the freshwater marsh in the Sanjiang Plain, Northeast China. Environmental Pollution and Public Health (EPPH2008), Shanghai; 2008.

Chapter 4
Dynamics of Dissolved Carbon and Nitrogen in Wetland Soil and its Overlying Ice and Water During the Freeze–Thaw Period

The freeze–thaw process can significantly affect soil physicochemical properties and microbial activity as well as elemental cycling. Dynamics of the concentrations of dissolved carbon and nitrogen in wetland soil during the freeze–thaw period comprehensively reflect these processes. The accumulation and release of wetland soil carbon and nitrogen during the freeze–thaw period will directly or indirectly affect the content of dissolved carbon and nitrogen in the overlying ice and water of the wetland. The effects of freeze–thaw on the soil microbial biomass and its composition, the concentrations of dissolved carbon and nitrogen in soil, and greenhouse gas emissions have been widely studied, but the results vary greatly and some have reached opposing conclusions (Henry 2007). The differences are mainly caused by different experimental methods. For example, the temperature and time of freeze–thaw, and the seasons in which soil samples were collected are likely to have a considerable impact on the experimental results. In addition, most of these previous studies were conducted using indoor simulation methods.

In this chapter, through semi-simulation and the direct use of field freeze–thaw conditions, the dynamics of DOC, NH_4^+–N and NO_3^-–N concentrations in wetland soil, and its overlying ice and water were studied during the spring freeze–thaw period. The results can contribute to a better understanding of the accumulation and release processes of wetland soil carbon and nitrogen during the freeze–thaw period.

4.1 Dynamics of Dissolved Carbon and Nitrogen in Wetland Soil During the Freeze–Thaw Period

4.1.1 Physicochemical Properties of Soil

Physicochemical properties of the sampled soils are shown in Table 4.1.

X. Yu, *Material Cycling of Wetland Soils Driven by Freeze–Thaw Effects*,
Springer Theses, DOI: 10.1007/978-3-642-34465-7_4,
© Springer-Verlag Berlin Heidelberg 2013

Table 4.1 Selected physical and chemical properties of soil samples ($n = 3$)

	H	M
Clay content (%)	40.2 ± 11.2	64.5 ± 8.09
pH	5.6 ± 0.5	5.4 ± 0.6
Organic matter content (%)	12.9 ± 3.13	11.7 ± 3.27
DOC content (mg/kg)	2762.00 ± 182.00	546.00 ± 76.20
NH_4^+–N (mg kg^{-1})	648.51 ± 102.31	871.74 ± 79.69
NO_3^-–N (mg kg^{-1})	1.28 ± 0.13	1.19 ± 0.10

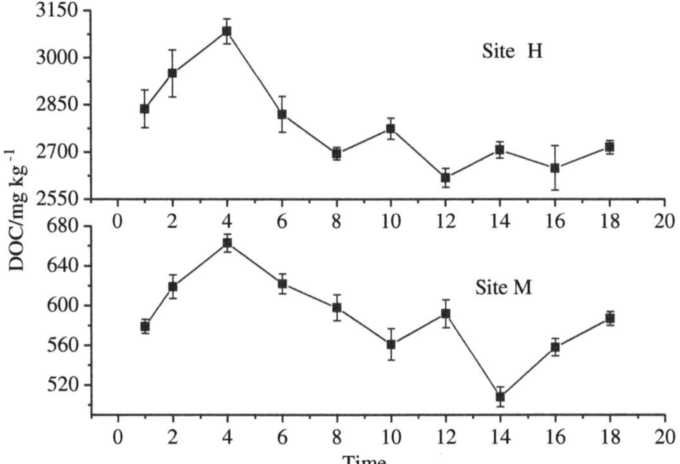

Fig. 4.1 Dynamics of DOC concentrations in wetland soils during freeze–thaw period

4.1.2 Dynamics of Dissolved Carbon and Nitrogen in Wetland Soil During the Freeze–Thaw Period

Dynamics of DOC concentrations during the freeze–thaw period in the *Carex lasiocarpa* wetland soil and *Calamagrostis angustifolia* wetland soil are shown in Fig. 4.1. At day 1 of the freeze–thaw period, DOC concentrations in the *Carex lasiocarpa* wetland soil and *Calamagrostis angustifolia* wetland soil were 2837 mg kg^{-1} and 579 mg kg^{-1}, respectively, which are 2.72 and 6.04 % higher than concentrations in the background soil. From day 1 to 4 of the freeze–thaw period, soil DOC concentrations in the *Carex lasiocarpa* and *Calamagrostis angustifolia* wetlands showed an increasing trend, and the soil DOC concentrations on the fourth day of the 18-day freeze–thaw period reached their maximum values of 3084 mg kg^{-1} and 663 mg kg^{-1} in the *Carex lasiocarpa* and *Calamagrostis angustifolia* wetlands, respectively. From day 5, DOC concentrations in the soil decreased with fluctuation. Soil DOC concentrations in the *Carex lasiocarpa* and *Calamagrostis angustifolia* wetlands were 2775 mg kg^{-1} and 587 mg kg^{-1}, respectively, on day 20 of the freeze–thaw period.

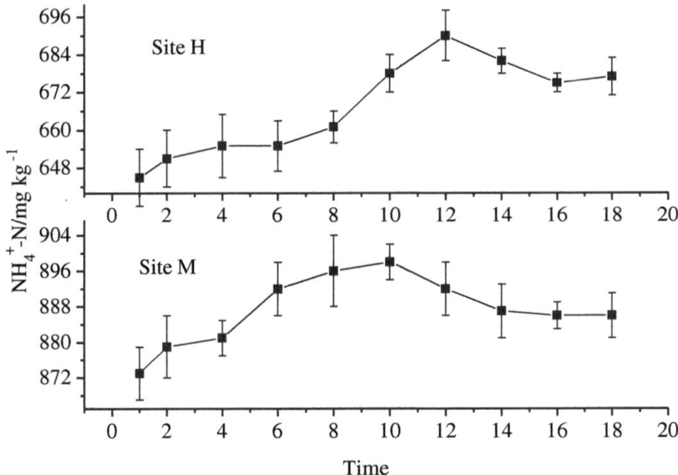

Fig. 4.2 Dynamics of NH_4^+-N concentrations in wetland soils during freeze–thaw period

In the first few days of the freeze–thaw period, soil DOC concentrations increased significantly in the *Carex lasiocarpa* and *Calamagrostis angustifolia* wetlands, presumably mainly because of the lethal effects of the freeze–thaw process on microorganisms. During the decomposition of dead microorganisms, some small molecular weight sugar and amino acids would be released, which would increase DOC concentrations in the soil. However, from day 4, DOC concentrations began to decline, indicating that multiple freeze–thaw cycles reduced DOC concentrations in the soil although the freeze–thaw promoted the release of DOC in the short term. This result is consistent with previous studies. For example Grogan and Michelsen (2004) showed that DOC concentrations in the soil reached a maximum after 1–3 cycles of freeze–thaw, and that DOC concentrations in the soil decreased after six cycles of freeze–thaw.

The dynamics of NH_4^+-N concentrations in the soils from the *Carex lasiocarpa* and *Calamagrostis angustifolia* wetlands during the freeze–thaw period are shown in Fig. 4.2. NH_4^+-N concentrations in the *Carex lasiocarpa* and *Calamagrostis angustifolia* wetland soils initially increased and then decreased over time. The soil NH_4^+-N concentration in the *Carex lasiocarpa* wetland reached a maximum of 690.32 mg kg^{-1} on the 12th day, and the soil NH_4^+-N concentration in the *Calamagrostis angustifolia* wetland reached its maximum of 898.14 mg kg^{-1} on the 10th day. After these maxima, soil NH_4^+-N concentrations gradually decreased in the *Carex lasiocarpa*, and *Calamagrostis angustifolia* wetlands, and were 677.31 and 886.45 mg kg^{-1}, respectively, on day 18 of the freeze–thaw period. These concentrations are 104.48 and 101.77 % of the soil background concentrations, respectively. Throughout the whole freeze–thaw period, NH_4^+-N concentrations were higher than that of the background, mainly due to the promoting impact of the freeze–thaw process on the mineralization of

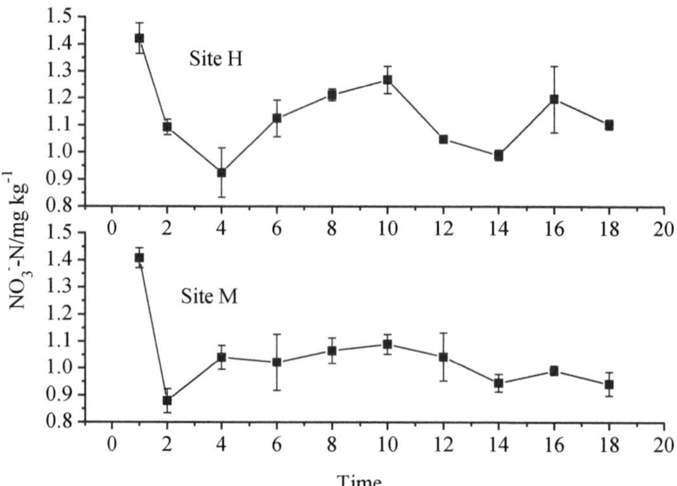

Fig. 4.3 Dynamics of NO^-_3–N concentrations in wetland soils during freeze–thaw period

soil organic nitrogen. In the early part of the freeze–thaw period, dead microorganisms presumably provided considerable matrices for other microorganisms and released large amounts of inorganic nitrogen, which was reflected by the significant increase in NH_4^+–N concentrations. However, with gradual consumption of the organic substrates, the influence of the freeze–thaw process on the mineralization of organic nitrogen decreased. In addition, nitrification may have consumed NH_4^+–N, leading to a gradual decrease in NH_4^+–N concentrations in the soil.

The dynamics of soil NO_3^-–N concentrations in the *Carex lasiocarpa* and *Calamagrostis angustifolia* wetlands during the freeze–thaw period are shown in Fig. 4.3. On day 1 of the freeze–thaw period, soil NO_3^-–N concentrations in the *Carex lasiocarpa* and *Calamagrostis angustifolia* wetlands were 1.42 and 1.41 mg kg^{-1}, respectively, which were 0.14 and 0.22 mg kg^{-1} higher than the soil background concentrations. From day 1 to 4, NO_3^-–N concentrations in the *Carex lasiocarpa* wetland soil showed a downward trend and were 1.12 mg kg^{-1} on day 4. In the first two days of the freeze–thaw period, the NO_3^-–N concentration in the *Calamagrostis angustifolia* wetland soil declined to some extent and was 1.02 mg kg^{-1} on the 2nd day. After that, NO_3^-–N concentrations in general remained stable with little fluctuation. On day 18 of the freeze–thaw period, NO_3^-–N concentrations in the *Carex lasiocarpa* and *Calamagrostis angustifolia* wetland soils were 1.10 and 0.94 mg kg^{-1}, respectively, which are lower than those of the background soil concentrations. In the first few days of the freeze–thaw period, NO_3^-–N concentrations in the *Carex lasiocarpa* and *Calamagrostis angustifolia* wetland soils were higher than the background soil concentrations, probably because of the promoting effects of the freeze–thaw on the mineralization of soil organic nitrogen and the release of inorganic nitrogen. However, the freeze–thaw cycle is also able to promote nitrification of soil.

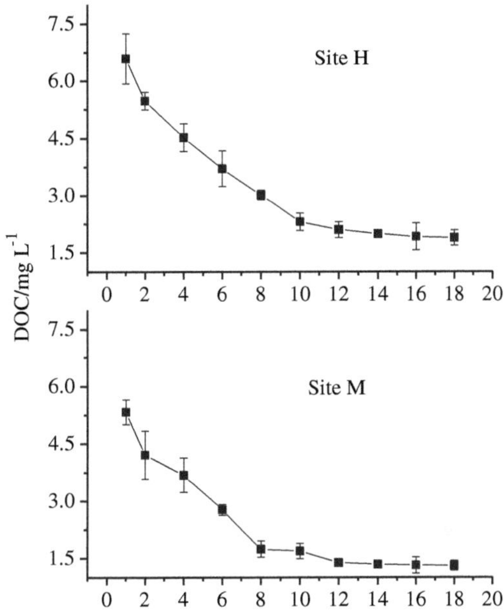

Fig. 4.4 Dynamics of DOC concentrations in ice during freeze–thaw period

When the increase in NO_3^--N concentration by the enhanced mineralization is less than the NO_3^--N consumed by enhanced nitrification, the NO_3^--N concentration will decrease. This is probably the main reason for the decrease or fluctuation of NO_3^--N concentrations in the *Carex lasiocarpa* and *Calamagrostis angustifolia* wetland soils.

4.2 Dynamics of Dissolved Carbon and Nitrogen in Overlying Ice of the Wetland During the Freeze–Thaw Period

Dynamics of DOC concentrations in the overlying ice of wetlands during the spring freeze–thaw period are shown in Fig. 4.4. The DOC concentration in the overlying ice of the *Carex lasiocarpa* wetland is higher than that of the *Calamagrostis angustifolia* wetland, and both of them show a downward trend with time. During the 18-day freeze–thaw period, the DOC concentrations in the overlying ice of the *Carex lasiocarpa* and *Calamagrostis angustifolia* wetlands decreased from 6.59 and 5.33 mg L^{-1} to 1.89 and 1.31 mg L^{-1}, respectively.

A first-order exponential decay equation was applied to fit DOC concentrations in the overlying ice of the *Carex lasiocarpa* and *Calamagrostis angustifolia* wetlands during the spring freeze–thaw period as:

$$DOC_t = DOC_0 e^{-kt}$$

Table 4.2 First-order decay parameters of DOC concentrations in wetland ice during spring freeze–thaw period

Site	DOC_0 (mg L^{-1})	k (d^{-1})	R^2
H	6.741	0.091	0.961
M	5.551	0.110	0.947

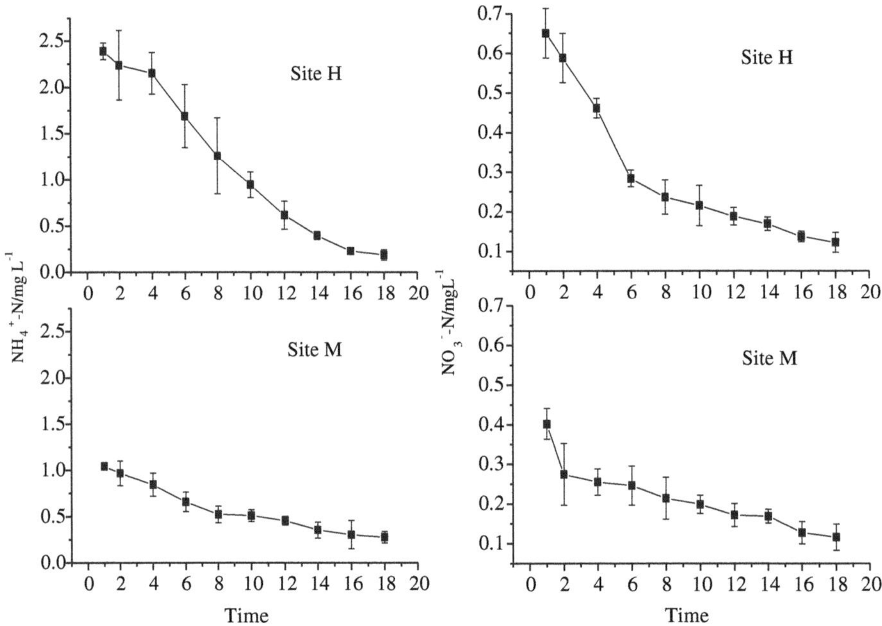

Fig. 4.5 Dynamics of NH$_4{}^+$–N and NO$^-{}_3$–N concentrations in ice during freeze–thaw period

where DOC$_t$ (mg L^{-1}) is the DOC concentration in the ice at time t (d), DOC$_0$ (mg L^{-1}) is the initial concentration of DOC in the ice, k (d^{-1}) is the rate attenuation constant of the DOC concentration, and t (d) is time. Values of various parameters in the equation are shown in Table 4.2. It can be seen that the first order exponential decay equation describes very well the changes in DOC concentrations in the overlying ice of the *Carex lasiocarpa* and *Calamagrostis angustifolia* wetlands over time ($R^2 > 0.947$).

Similarly to the trend of DOC concentration change, concentrations of NH$_4{}^+$–N and NO$_3{}^-$–N in the ice during the freeze–thaw period showed a decreasing trend over time (Figure 4.5). In the 18-day freeze–thaw period, NH$_4{}^+$–N and NO$_3{}^-$–N concentrations in the overlying ice decreased by 2.202 and 0.528 mg L^{-1}, respectively in the *Carex lasiocarpa* wetland, and decreased by 0.763 and 0.286 mg L^{-1}, respectively in the *Calamagrostis angustifolia* wetland. At the beginning of the freeze–thaw period, the NH$_4{}^+$–N concentration in the overlying ice of the *Carex lasiocarpa* wetland was 1.349 mg L^{-1} higher than that of the *Calamagrostis*

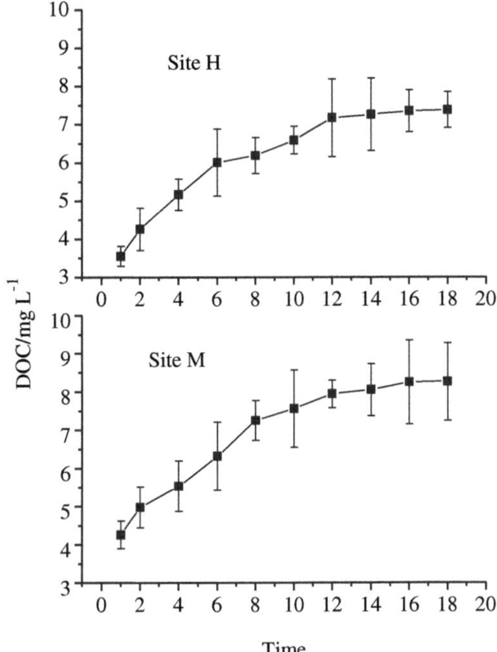

Fig. 4.6 Dynamics of DOC concentrations in water during freeze–thaw period

angustifolia wetland, but it was 0.090 mg L^{-1} lower than that of the *Calamagrostis angustifolia* wetland after 18 days of the freeze–thaw period. At the beginning of the freeze–thaw period, the NO$_3^-$–N concentration in the overlying ice of the *Carex lasiocarpa* wetland was 0.248 mg L^{-1} higher than that of the *Calamagrostis angustifolia* wetland, but it was 0.006 mg L^{-1} lower than that of the *Calamagrostis angustifolia* wetland after 18 days of the freeze–thaw period.

4.3 Dynamics of the Dissolved Carbon and Nitrogen in Overlying Water of the Wetland During the Freeze–Thaw Period

Figure 4.6 shows the dynamics of DOC concentrations in overlying water of the *Carex lasiocarpa* and *Calamagrostis angustifolia* wetlands during the freeze–thaw period. During the 18-day freeze–thaw period, DOC concentrations in the overlying water showed an upward trend over time, and DOC concentrations in the overlying water of the *Carex lasiocarpa* wetland increased from 4.26 to 8.26 mg L^{-1}, and that in the *Calamagrostis angustifolia* wetland increased from 3.55 to 7.38 mg L^{-1}, which was slightly higher than the DOC concentration in the wetlands' overlying water of the previous year (*Carex lasiocarpa* wetland: 7.17 mg L^{-1};

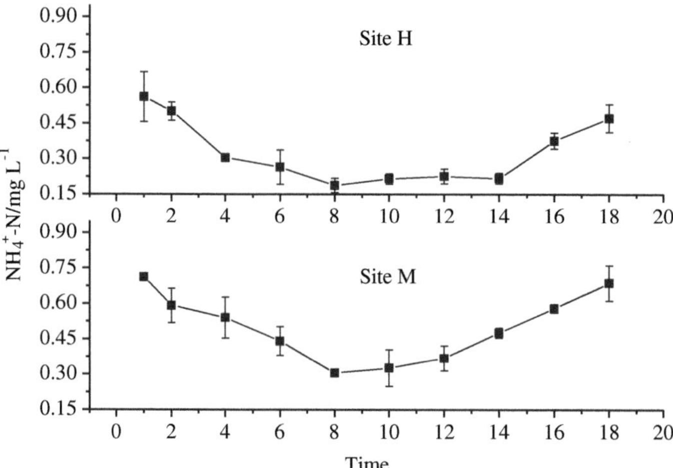

Fig. 4.7 Dynamics of $NO_3{}^-$–N concentrations in wetland water during freeze–thaw period

Calamagrostis angustifolia wetland: 6.23 mg L^{-1}). During the first 10 days of the freeze–thaw period, DOC concentrations in the overlying water of the *Carex lasiocarpa* wetland significantly increased, and DOC concentrations in the overlying water of the *Carex lasiocarpa* and *Calamagrostis angustifolia* wetlands were relatively stable from the 14th to 18th day of the freeze–thaw period. A certain amount of DOC was released from the melt of overlying ice into the overlying water in wetlands during the freeze–thaw period. Meanwhile there was distribution equilibrium of DOC between the overlying water and soil. DOC in the wetland soil can be released into the overlying water through diffusion and other physicochemical processes, and DOC in the overlying water could enter the soil through adsorption and diffusion processes. The DOC concentration in the overlying water is closely related to the soil and microbial activity at the same time. The dynamics of DOC concentrations in the overlying water of wetlands during the freeze–thaw period comprehensively reflects these processes.

Figure 4.7 shows the dynamics of $NH_4{}^+$–N concentrations in the overlying water of the *Carex lasiocarpa* and *Calamagrostis angustifolia* wetlands during the freeze–thaw period. From day 1 to 8 of the freeze–thaw period, $NH_4{}^+$–N concentrations in the overlying water of the *Carex lasiocarpa* and *Calamagrostis angustifolia* wetlands decreased from 0.562 and 0.712 mg L^{-1} to 0.187 and 0.306 mg L^{-1}. From day 8 to 14, $NH_4{}^+$–N concentrations in the overlying water of the *Carex lasiocarpa* wetland were relatively stable, and $NH_4{}^+$–N concentrations in the overlying water of the *Calamagrostis angustifolia* wetland were relatively stable from day 8 to 12. $NH_4{}^+$–N concentrations in the overlying water of the *Carex lasiocarpa* wetland began to rise from the 14th day and reached 0.472 mg L^{-1} on the 18th day; and $NH_4{}^+$–N concentrations in the overlying water of the *Calamagrostis angustifolia* wetland began to rise from the 12th day and reached 0.687 mg L^{-1} on the 18th day.

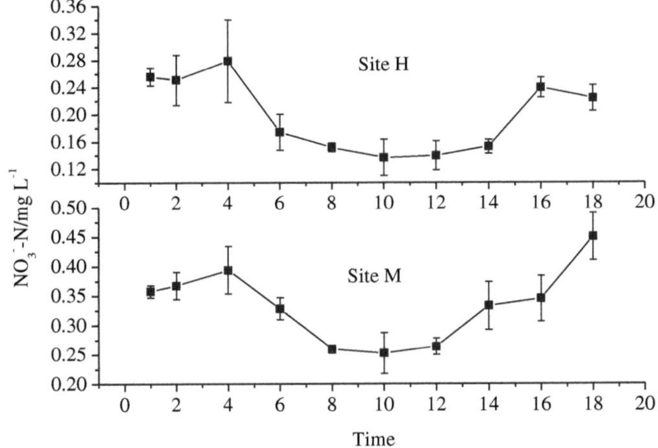

Fig. 4.8 Dynamics of NO$_3^-$–N concentrations in wetland water during freeze–thaw period

Figure 4.8 shows the dynamics of NO$_3^-$–N concentrations in the overlying water of the *Carex lasiocarpa* and *Calamagrostis angustifolia* wetlands during the freeze–thaw period. From day 1 to 4 of the freeze–thaw period, NO$_3^-$–N concentrations in the overlying water of the *Carex lasiocarpa* and *Calamagrostis angustifolia* wetlands increased from 0.256 and 0.358 mg L^{-1} to 0.279 and 0.394 mg L^{-1}, respectively. From day 4 to 8, NO$_3^-$–N concentrations in the overlying water of the *Carex lasiocarpa* and *Calamagrostis angustifolia* wetlands showed a declining trend, and reached 0.152 and 0.259 mg L^{-1}, respectively on day 8. NO$_3^-$–N concentrations in the overlying water of the *Carex lasiocarpa* and *Calamagrostis angustifolia* wetlands were stable from day 8 to 14 and from day 8 to 12, respectively. Concentrations subsequently increased and reached 0.224 and 0.450 mg L^{-1}, respectively on day 18 of the freeze–thaw period.

4.4 Chapter Summary

In this chapter, through field semi-simulation, the dynamics of DOC, NH$_4^+$–N and NO$_3^-$–N concentrations in the *Carex lasiocarpa* and *Calamagrostis angustifolia* wetland soils, and its overlying ice and water were studied during the spring freeze–thaw period. DOC and NH$_4^+$–N concentrations in the wetland soil during the freeze–thaw period first increased and then decreased. NO$_3^-$–N concentrations initially decreased and then tended to be stable. DOC, NH$_4^+$–N and NO$_3^-$–N concentrations in the overlying ice of the wetlands all showed downward trends, while DOC concentrations in the water overlying wetlands showed an increasing trend; and NH$_4^+$–N and NO$_3^-$–N concentrations showed decreasing trends first and then increased. The results should contribute to a better understanding of the accumulation and release of wetland soil carbon and nitrogen during the freeze–thaw period.

References

Grogan P, Michelsen A, Ambus P, Jonasson S. Freeze–thaw regime effects on carbon and nitrogen dynamics in sub–arctic heath tundra mesocosms. Soil Biol Biochem. 2004;36:641–54.

Henry HAL. Soil freeze–thaw cycle experiments: trends, methodological weaknesses and suggested improvements. Soil Biol Biochem. 2007;39:977–86.

Chapter 5
Freeze–Thaw Effects on Sorption/ Desorption of Dissolved Organic Carbon in Wetland Soils

Dissolved organic carbon (DOC) is a major controlling factor in soil formation (Dawson et al. 1978), mineral weathering (Raulund–Rasmussen et al. 1998), nutrient cycling, microbial activity, and organic matter decomposition and transformation in soils (Magill and Aber 2000; Williams et al. 2000). The sorption of DOC is the dominant factor in DOC concentration in soil solutions, transport and transformation and the microbial availability of DOC. It is not only the major control for organic matter (OM) and OM–assisted transport (Kaiser and Zech 1999); it also contributes to the stabilization and accumulation of organic matter in soils (Guggenberger and Kaiser 2003).

Six mechanisms have been suggested to be involved in the sorption of DOC to soil mineral surfaces: ligand exchange, cation bridges, anion exchange, cation exchange, van der Waals interactions and hydrophobic effects (Jardine et al. 1989). The ligand exchange between carboxyl/hydroxyl functional groups of dissolved organic matter (DOM) and iron oxide surfaces was the dominant interaction mechanism, especially under acidic or slightly acidic conditions (Gu et al. 1994). Some researchers investigated the important factors influencing the sorption of DOC in soils. The sorption capacity of DOC in clay soils was greater than that in sandy soils (Moore and Matos 1999), which appeared to be positively correlated to the soil clay content (Shen 1999). Moore et al. 1992 observed that the sorption of DOC was related to soil organic carbon content (positive) and oxalate–extractable Al and dithionite–extractable Fe (negative). However, the results of another research indicated that soils with higher organic matter content adsorbed less DOC (Jardine et al. 1989). And the sorption of DOC in acid soils was greater than that in alkaline soils (Moore et al. 1992). Maximum sorption of DOC occurred at pH 4, and decreased with either an increase or decrease in pH (Ussiri and Johnson 2004). Lilienfein et al. (2004) elucidated that the sorption of DOC increased significantly with increasing soil development, whereas the soil depth did not have a consistent and significant effect. In addition, Vandenbruwane et al. (2007) did the comparison of different isotherm models for dissolved organic carbon. Many studies about sorption of

X. Yu, *Material Cycling of Wetland Soils Driven by Freeze–Thaw Effects*,
Springer Theses, DOI: 10.1007/978-3-642-34465-7_5,
© Springer-Verlag Berlin Heidelberg 2013

DOC in soils have been documented, all of which have focused on farmland soils and forest soils, moreover little attention has been paid on the sorption/desorption of DOC in wetland soils. Moreover the studies on DOC sorption potential of soils have been conducted in a mild temperature range from 5 to 25 °C condition soils. Little is available on the sorption/desorption of DOC in freezing and thawing soils.

Wetlands are lands transitional between terrestrial and aquatic systems where the water table is usually at or near the surface or the land is covered by shallow water (Mitsch and Gosselink 2000). Due to different properties between wetland soils and other soils, such as hydroperiod, the sorption/desorption behavior of DOC likely differs, the study results of DOC sorption in agricultural and forest soils do not necessarily fit DOC sorption in wetland soils. DOC can transport from soil solution to overlying water column through the process of molecular diffusion (Arnarson and Keil 2000); and sorption/desorption is the major process of DOC partition between soil mineral surfaces and soil solution. Therefore, the research on DOC sorption in wetland soils is necessary. In addition, many wetlands have been reclaimed to farmlands in China, especially in Sanjiang Plain of Northeast China, where the large–scale freshwater marshes have been converted to agricultural areas. The effect of wetland reclamation on the DOC sorption/desorption is not examined yet, which might alter the loss of carbon in cultivated wetlands.

Furthermore, for wetlands in the cold region, freeze–thaw is an important environmental characteristic. Maehlum et al. (1995) reported that cold winter climate can affect hydraulic processes and biogeochemical processes of wetlands. Also, freeze–thaw cycles have effects on soil physical properties, microbial activity and microbial community composition (Schadt et al. 2003; Lipson and Schmidt 2004; Six et al. 2004; Sjursen et al. 2005). Freeze–thaw cycles have disruptive effects on soils structure, which decrease bulk density and penetration resistance (Unger 1991). Likewise, in controlled laboratory incubations, freeze–thaw cycles resulted in decreases of soil aggregate stability, particularly at high soil moisture (Oztas and Fayetorbay 2003). Wang et al. (2007) have studied the freeze–thaw effects on phosphorus sorption in wetland soils. In their study results, freeze–thaw can promote phosphorus sorption in wetland soils and these effects accumulate over successive freeze–thaw cycles (ranging from one to six cycles). However, whether and how freeze–thaw affects sorption/desorption of DOC in wetland soils is still unknown. Moreover, there has been growing interest in how changes in soil freezing initiated by climate change might alter soil nutrient dynamics (Henry 2007). Thus whether global warming can affect sorption/desorption of DOC through affecting freeze–thaw cycles also needs to be concerned.

The objective of this study was to investigate freeze–thaw effects on sorption/desorption of DOC in wetland soils and reclaimed wetland soils. Particular attention was given to possible effects of reclamation and global warming on the retention and release of DOC in wetland soils. The understanding of DOC sorption/desorption in the cold region, therefore, can be verified with greater confidence level based on the global–scale investigation.

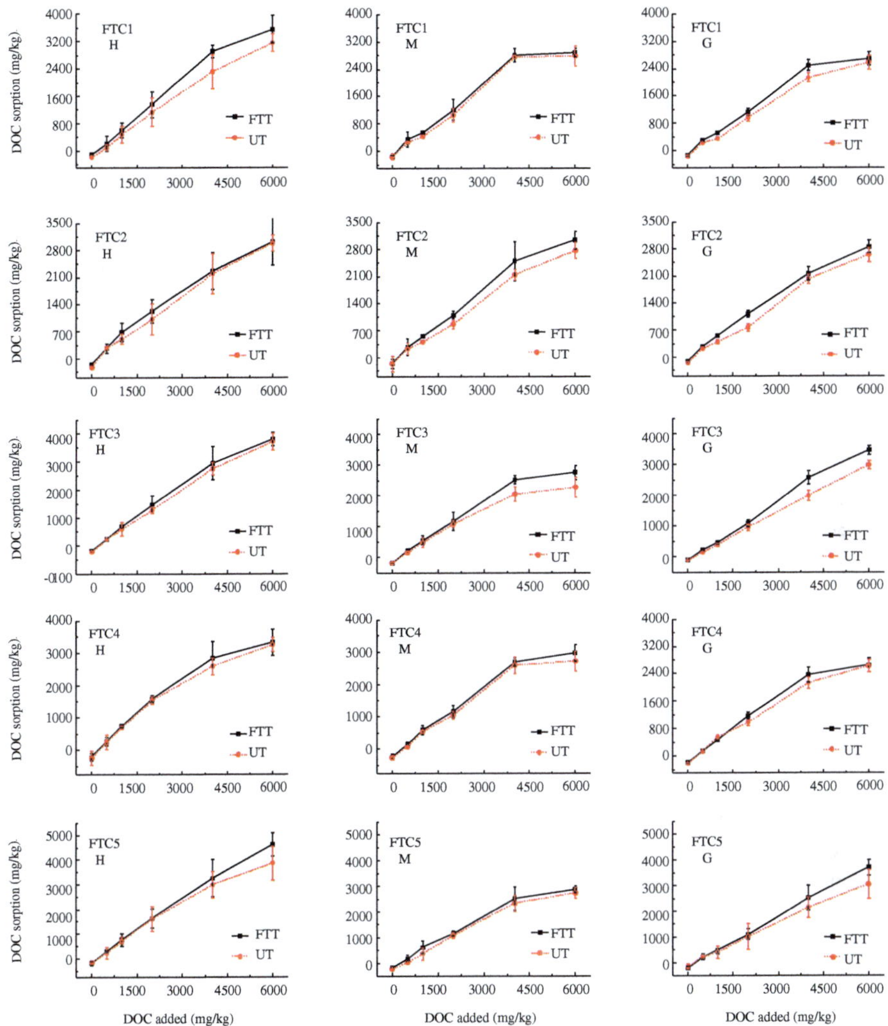

Fig. 5.1 Initial mass (IM) isotherms for DOC sorption in freeze–thaw treated wetland soils (With kind permission from Springer Science + Business Media: Chinese Geographical Science, Yu et al. (2010), Fig. 2, and any original (first) copyright notice displayed with material) *Solid lines* and *dashed lines* represent sorption of DOC in soils with and without freeze–thaw treatment, respectively. FTT and UT represent soils with and without freeze–thaw treatment, respectively. H and M represent two kinds of wetland soils. G represents farmland soils reclaimed from wetland. DOC added is identified on the x–axis. Error bars represent the standard error of the mean of three parallel samples

5.1 Sorption Behavior of DOC

The sorption amount of DOC was increased with DOC concentration added. This trend could be described by Initial Mass (IM) isotherm model, which is the most common model based on simple partitioning. This model is based on the linear adsorption isotherm (Travis and Etnier 1981), but additionally accounts for the substances initially present within the soils:

$$\text{IM isotherm}: \quad S = m \times C_1 - b$$

Where S is the total amount of DOC adsorbed (mg/kg), C_1 is the concentration of DOC added (mg/kg), m is the partition coefficient of DOC between soils and solution, used to estimate the sorption capacity, b is the intercept (mg/kg).

IM isotherms for sorption of DOC in FTT soils and UT soils were given in Fig. 5.1. Parameters of the model were presented in Table 5.1. IM isotherm was used to describe sorption of DOC in soils in many studies (Nodvin et al. 1986; Guggenberger and Zech 1992; Riffaldi et al. 1998), and the highest concentration of DOC added in experiment was 81 mg/L (Nodvin et al. 1986). However, in our study, for M and G soils, when DOC concentration was added from 0 to 400 mg/L, IM isotherm can describe sorption behaviour of DOC very well. When DOC added was higher than 400 mg/L, IM isotherm was not suitable. Moreover, for H soils, the highest concentration of 600 mg/L was still suitable to IM isotherm.

5.2 Freeze–Thaw Effects on Sorption of DOC

In Fig. 5.1, the sorption isotherms of FTT soils were always higher than that of UT soils. With a nonparametric test, the difference of the adsorbed DOC amount between the FTT soils and UT soils was significant ($Z = -7.784$; $p < 0.001$). These results indicated that freeze–thaw cycles increased the adsorbed amount of DOC in soils. Moreover, in Table 5.1, values of m, used to estimate the sorption capacity, of FTT soils were always higher than that of UT soils. With a nonparametric test, the difference of m values between FTT soils and UT soils was also significant ($Z = -3.408$; $p = 0.001$). These results indicated that freeze–thaw increased the sorption capacity of DOC in soils, which might be resulted from the freeze–thaw effects on the soil structures. The main clay mineral of these soils is hydrous mica. Freezing could break down aggregates in soils, and soil mineral surfaces might be increased, which might provide more sorption positions for DOC (Hinman 1970; Bullock et al. 1988; Oztas and Fayetorbay 2003; Six et al. 2004). In addition, DOC sorption was affected by iron oxides in soils. About 50–70 % DOC was adsorbed by oxides in soils (Jardine et al. 1989). We speculated that the amorphous iron oxides in soils could be increased by alternate freeze–thaw cycles. The increased iron oxides amount could affect DOC sorption in soils. At present, the further research about freeze–thaw effects on iron oxides in wetland soils is needed. And the results will confirm this speculation. Moreover, the increased sorption capacity of DOC caused by freeze–thaw cycles could increase DOC retention in wetland soils. Under conditions of low plant and microbial activity in the winter, the sorption of DOC may represent an intermediate DOC sink in soil. Thus, the increased DOC retention in wetland

Table 5.1 Initial mass isotherms parameters for DOC sorption in freeze–thaw treated wetland soils

		H		M		G	
Parameters	Cycles	FTT	UT	FTT	UT	FTT	UT
m	FTC1	0.758	0.628	0.725	0.718	0.646	0.562
	FTC2	0.590	0.580	0.649	0.560	0.556	0.522
	FTC3	0.776	0.729	0.664	0.553	0.668	0.525
	FTC4	0.754	0.697	0.715	0.702	0.640	0.570
	FTC5	0.856	0.793	0.666	0.650	0.648	0.534
b	FTC1	142	170	142	211	114	145
	FTC2	22	114	123	143	35.7	135
	FTC3	126	160	134	111	147	112
	FTC4	83	72	199	251	169	142
	FTC5	132	101	138	259	150	89
R	FTC1	0.999	0.999	0.997	0.995	0.999	0.997
	FTC2	0.994	0.996	0.999	0.996	0.998	0.994
	FTC3	0.999	0.999	0.999	0.997	0.998	1.000
	FTC4	0.995	0.990	0.999	0.998	0.999	0.997
	FTC5	0.999	0.998	0.999	0.999	0.999	0.999

Note FTT and UT represent soils with and without freeze–thaw treatment, respectively. FTCs represent freeze–thaw cycles. m is partition coefficient of DOC between soils and solution. b is the intercept of Initial mass isotherm model. Values of m, b and R are obtained from linear regression analysis. Values of m and b were calculated from IM equation when the adsorbed DOC amount and the added DOC amount were put into the equation

soils by freeze–thaw cycles likely increase the DOC sink in wetland soils. For DOC sorption could decrease the bioavailability of DOC, freeze–thaw could have effects on the microbial contacting to the DOC in the following growing season.

Furthermore, the difference of the adsorbed DOC amount at varied FTCs (from FTC1 to FTC5) was not significant (H soils: $F = 0.43$, $p = 0.996$; M soils: $F = 0.05$, $p = 1.000$; G soils: $F = 0.31$, $p = 0.998$). The results indicated that the FTCs could not significantly affect the adsorbed DOC amount in soils. The difference of m values was not significant at varied FTCs (H soils: $R = 0.588$, $p = 0.298$; M soils: $R = 0.243$, $p = 0.294$; G soils: $R = 0.319$, $p = 0.600$), which indicated that FTCs did not have significant effects on the sorption capacity of DOC. Therefore, the first FTC increased the DOC sorption in soils, while the following FTCs on DOC sorption did not increase or reduce these effects.

Comparison of adsorbed DOC amount among varied soils at sites H, M and G was presented (Fig. 5.1). H soils adsorbed more DOC than M and G soils did (H > M > G). The m values of soils at site H were about 11.8 % higher than that at site M, and were about 18.2 % higher than that at site G. The m values of soils at site M were about 8.3 % higher than that at site G. These results indicated that the DOC sorption capacity of soils at site H was greater than site M and G, while

Table 5.2 Soil physical and chemical properties of sampling sites in Sanjiang Plain, Northeast China

	H	M	G
Clay content (%)	40.20 ± 11.20	64.50 ± 8.09	46.80 ± 4.12
pH	5.60 ± 0.50	5.40 ± 0.60	6.00 ± 0.50
Organic matter content (%)	12.90 ± 3.13	11.70 ± 3.27	13.30 ± 3.09
DOC content (mg/kg)	2762.00 ± 182.00	546.00 ± 76.20	197.00 ± 31.50

Values are given as mean ± standard error, n = 3

the DOC sorption capacity of soils at site M was greater than that at site G. It was reported that soils with higher organic matter content adsorbed less DOC (Jardine et al. 1989). And there was positive correlation between soil sorption capacity of DOC and the soil clay content (Shen 1999). Consistent with these previous studies, the adsorbed amount and sorption capacity of DOC in soils at site M were greater than that at site G. The soil clay content at site M was about 37.8 % higher than that at site G, while the soil organic matter content at site G was about 13.7 % higher than that at site M (Table 5.2). On contrast, the adsorbed amount of DOC at site H was greater than that at site M and G, although the soil organic matter content at site H was greater than that at site M and G. It was not consistent with the previous studies. Thus another factor might exist for the greater sorption capacity of H soils. In fact, a certain content of peat exists in the surface soils at site H. This is a ubiquitous phenomenon for perennial flooded wetland soils. The peat is porous and has great sorption capacity. This might be the reason for H soils with higher organic matter content and lower clay content adsorbing more DOC than M and G soils.

5.3 Freeze–Thaw Effects on Desorption of DOC

After the sorption experiment, some soil samples were reused for desorption experiment of DOC. DOC was desorbed from soil samples which had adsorbed DOC in the sorption experiment. The desorption potential of DOC was estimated by the percentage of the desorbed DOC amount to the adsorbed DOC amount (PDS). PDS in FTT and UT soils were presented in Table 5.3. With a nonparametric test, there was significant difference of PDS in FTT and UT soils ($Z = -4.082$; $p < 0.001$). PDS of FTT soils was lower than that of UT soils, which might indicate that freeze–thaw reduced the desorption potential of DOC. The comparison of PDS at varied sampling sites indicated that PDS order was G > M > H (Table 5.3). For example, after FTC5, PDS of H, M and G soils (with DOC concentration of 4000 mg/kg added in sorption experiment) was 31.7, 33.3 and 45.0 %, respectively.

Moreover, the difference of PDS in UT soils and FTT soils was increased with increasing FTCs (Fig. 5.2). It might indicated that the freeze–thaw effects on

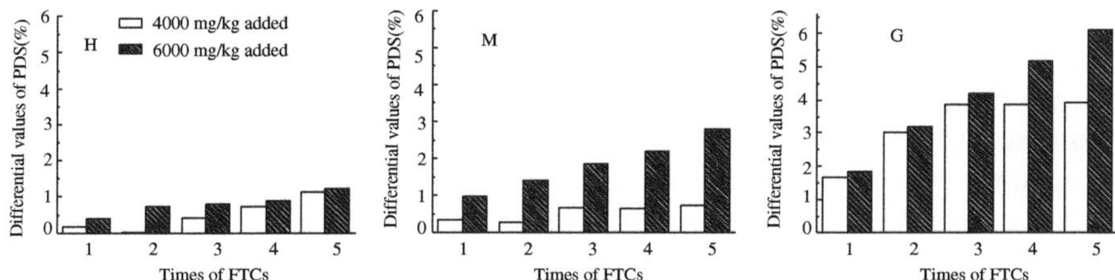

Fig. 5.2 Differential values of percentage of the desorbed DOC amount to adsorbed DOC amount (PDS) in freeze–thaw treated wetland soils. PDS represent percentage of desorbed DOC amount to adsorbed DOC amount in soils. H and M represented two kinds of wetland soils. G represents farmland soils reclaimed from wetland. The samples (containing 4000 and 6000 mg/kg added) after FTC5 in sorption experiment were selected to use in desorption experiment. FTCs are represented on x–axis. (With kind permission from Springer Science + Business Media: Chinese Geographical Science, Yu et al. (2010), Fig. 3, and any original (first) copyright notice displayed with material)

Table 5.3 The percentage of total desorbed amount of DOC in the desorption experiment to the adsorbed amount of NH_4^+ in the sorption experiment (%)

Soils	Cycles	Samples with 4000 mg/kg DOC added in sorption experiment		Samples with 6000 mg/kg DOC added in sorption experiment	
		FTT	UT	FTT	UT
H soils	FTC1	8.8	9.2	8.0	8.2
	FTC2	15.6	16.3	14.9	14.9
	FTC3	21.3	22.1	19.4	19.8
	FTC4	26.7	27.6	23.6	24.3
	FTC5	31.7	32.9	27.3	28.5
M soils	FTC1	9.2	9.6	9.7	10.7
	FTC2	17.3	18.0	17.3	18.7
	FTC3	23.0	23.7	23.5	25.4
	FTC4	28.3	28.9	29.0	31.2
	FTC5	33.3	33.6	34.1	36.9
G soils	FTC1	12.1	13.8	9.8	11.6
	FTC2	23.0	26.0	17.9	21.1
	FTC3	30.9	34.8	23.9	28.1
	FTC4	38.1	41.9	28.9	34.1
	FTC5	44.9	48.7	33.8	39.9

Note FTT and UT represent soils with and without freeze–thaw treatment, respectively. FTCs represent freeze–thaw cycles

DOC desorption could be improved by the increasing FTCs, which meant that the decreased desorption of DOC caused by freeze–thaw treatment were increased with FTCs. Moreover, in the cold region, the annual average FTCs were stable. The freeze–thaw effects on the DOC desorption potential might also be stable. If air temperature increased in the cold region and the FTCs reduced, the DOC desorption potential might increase because of the reduced effects of freeze–thaw. The increased desorption could improve the release of DOC. Therefore, DOC release from soil mineral surfaces to soil solution could be increased by global warming. And DOC transporting from soil solution to overlying water column through the process of molecular diffusion might be affected, which might affect DOC concentration in soils and the overlying water column. However, besides reducing FTCs, global warming might also affect freeze–thaw time, freeze–thaw intensity, and so on. Whether these factors response to global warming could affect sorption/desorption of DOC needs further investigation.

5.4 Effects of Wetland Reclamation on the Soil Sorption/ Desorption Behaviors of DOC

The comparisons of the DOC sorption and desorption behaviours in soils at varied sites (Figs. 5.1 and 5.2) indicated that the DOC sorption capacity in soils at site G was lower than that at site M, and the DOC desorption potential was greater than that at site G. However, the soil type of site G was the same as that of site M before the reclamation for soybean production. This indicated that the reclamation appears to reduce the sorption capacity of DOC and to increase the desorption potential of DOC. The iron oxides complex was decreased after the reclamation (Zou et al. 2008). The decreased iron oxides complex might reduce the sorption capacity and increase the desorption potential of DOC. The altered hydroperiod could affect amorphous iron oxides, which might be another reason. Su et al. (2001) observed that water logging increased amorphous iron oxides of soils. The increased amorphous iron oxides could result in the increased sorption of DOC. Due to the conversion of the wetlands to the farmlands, shortened hydroperiod in farmland might cause decreased amorphous iron oxides of soils and thus reduced sorption capacity of DOC in farmlands. Under the cultivation condition, lots of crop residues and other organic matters were returned to soils, which increased the organic matter content in soils. This resulted in the decreased sorption of DOC in the reclaimed wetland soils (Jardine et al. 1989). In addition, the decreased retention of DOC might increase the loss of carbon and the emission of greenhouse gas.

5.5 Chapter Summary

Freeze–thaw cycles can increase the sorption capacity of DOC and reduce desorption potential of DOC in wetland soils and reclaimed wetland soils. The freeze–thaw effects on the desorption potential of DOC can be improved by increasing FTCs. The wetland reclamation reduces the sorption capacity of DOC and increases the desorption potential of DOC.

References

Arnarson TS, Keil RG. Mechanisms of pore water organic matter adsorption to montmorillonite. Mar Chem. 2000;71:309–20.

Bullock MS, Kemper WD, Nelson SD. Soil cohesion as affected by freezing, water content, time and tillage. Soil Sci Soc Am J. 1988;52:770–6.

Dawson HJ, Ugolini FC, Hrutfiord BF, Zachara J. Role of soluble organics in the soil processes of a podzol, Central Cascades, Washington. Soil Sci. 1978;126:290–6.

Gu BH, Schmitt J, Chen ZH, Liang XY, McCarthy JF. Adsorption and desorption of natural organic matter on iron–oxide—mechanisms and models. Environ Sci Technol. 1994;28:38–46.

Guggenberger G, Zech W. Retention of dissolved organic carbon and sulfate in aggregated acid forest soils. J Environ Qual. 1992;21:643–53.

Guggenberger G, Kaiser L. Dissolved organic matter in soil: challenging the paradigm of sorptive preservation. Geoderma. 2003;113:293–310.

Henry AL. Soil freeze–thaw cycle experiments: trends, methodological weaknesses and suggested improvements. Soil Biol Biochem. 2007;39:977–86.

Hinman WG. Effects of freezing and thawing on some chemical properties of three soils. Can J Soil Sci. 1970;50:179–82.

Jardine PM, Weber NL, McCarthy JF. Mechanisms of dissolved organic carbon adsorption on soil. Soil Sci Soc Am J. 1989;53:1378–85.

Kaiser K, Zech W. Release of natural organic matter sorbed to oxides and a subsoil. Soil Sci Soc Am J. 1999;63:1157–66.

Lilienfein J, Qualls RG, Uselman SM, Bridgham SD. Adsorption of dissolved organic and inorganic phosphorus in soils of a weathering chronosequence. Soil Sci Soc Am J. 2004;68:620–8.

Lipson DA, Schmidt SK. Seasonal changes in an alpine soil bacterial community in the Colorado Rocky Mountains. Appl Environ Microbiol. 2004;70:2867–79.

Maehlum T, Jenssen PD, Warne WS. Cold-climate constructed wetlands. Water Sci Technol. 1995;32:95–101.

Magill AH, Aber JD. Variation in soil net mineralization rates with dissolved organic carbon addition. Soil Biol Biochem. 2000;32:597–601.

Mitsch WJ, Gosselink JG. Wetlands 3rd ed. New York: Wiley; 2000.

Moore TR, Souza WD, Koprivnjak JF. Controls on the sorption of dissolved organic carbon by soils. Soil Sci. 1992;2:120–9.

Moore TR, Matos L. The influence of source on the sorption of dissolved organic carbon by soils. Can J Soil Sci. 1999;79:321–4.

Nodvin SC, Driscoll CT, Likens GE. Simple partitioning of anions and dissolved organic carbon in a forest soil. Soil Sci. 1986;143:27–34.

Oztas T, Fayetorbay F. Effect of freezing and thawing processes on soil aggregate stability. Catena. 2003;52:1–8.

Raulund–Rasmussen K, Borrggaard OK, Hansen HCB, Olsson M. Effect of natural soil solutes on weathering rates of soil minerals. Eur J Soil Sci. 1998;49:397–406.

Riffaldi R, Levi–Minzi R, Saviozzi A. Adsorption on soil of dissolved organic carbon from farmyard manure. Agric Ecosyst Environ. 1998;69:113–9.

Schadt CW, Martin AP, Lipson DA, et al. Seasonal dynamics of previously unknown fungal lineages in tundra soils. Science. 2003;301:1359–61.

Shen YH. Sorption of natural dissolved organic matter on soil. Chemosphere. 1999;38:1505–15.

Six J, Bossuyt H, Degryse S, et al. A history of research on the link between (micro) aggregates, soil biota, and soil organic matter dynamics. Soil Tillage Res. 2004;79:7–31.

Sjursen H, Michelsen A, Holmstrup M. Effects of freeze–thaw cycles on microarthropods and nutrient availability in a sub-arctic soil. Appl Soil Ecol. 2005;28:79–93.

Su L, Zhang YS, Lin XY. Change of iron shape and phosphorus adsorption in paddy soils during alternate drying and wetting process. Plant Nutr Fertil Sci. 2001;7:410–5. in Chinese.

Travis CC, Etnier EL. A survey of sorption relationships for reactive solutes in soil. J Environ Qual. 1981;10:8–17.

Unger PW. Overwinter changes in physical–properties of no–tillage soil. Soil Sci Soc Am J. 1991;55:778–82.

Ussiri DAN, Johnson CE. Sorption of organic carbon fractions by Spodosol mineral horizons. Soil Sci Soc Am J. 2004;68:253–62.

Vandenbruwane J, De Neve S, Qualls RG, Sleutel S, Hofman G. Comparison of different isotherm models for dissolved organic carbon (DOC) and nitrogen (DON) sorption to mineral soil. Geoderma. 2007;139:144–53.

Wang G, Liu J, Zhao H, Wang J, Yu J. Phosphorus sorption by freeze–thaw treated wetland soils derived from a winter-cold zone. Geoderma. 2007;138:153–61.

Williams CF, Agassi M, Letey J. Facilitated transport of napropamide by dissolve organic matter through soil columns. Soil Sci Soc Am J. 2000;64:590–4.

Yu XF, Zhang YX, Zhao HM, et al. Freeze–thaw effects on sorption/desorption of dissolved organic carbon in wetland soils. Chin Geogr Sci. 2010;20(3):209–17.

Zou YC, Lu XG, Jiang M. Characteristics of the wetland soil iron under different ages of reclamation. Environ Sci. 2008;29(3):814–8. in Chinese.

Chapter 6
Freeze–Thaw Effects on Sorption/Desorption of Ammonium in Wetland Soils

Ammonium ions (NH_4^+) are a primary form of mineralized nitrogen (N) in most flooded wetland soils. NH_4^+ ions can be adsorbed by plants or be electrostatically held on negatively charged surfaces of soil particles. They can be prevented from further oxidation and are stable in wetland soils because of their anaerobic nature (Patrick and Reddy 1976). NH_4^+ sorption and desorption is an important mechanism of N retention and release in wetland soils that affects N availability for plant uptake and N leaching (Avnimelech and Laher 1977; Fenn et al. 1982; Phillips 1999).

NH_4^+ sorption in agricultural and forest soils has been extensively investigated (Lumbanraja and Evangelou 1994; Thompson and Blackmer 1992; Wang and Alva 2000). Soil properties, such as cation exchange capacity (CEC) and base saturation, affect NH_4^+ sorption in forest soils (Matschonat and Matzner 1996). The impact of soil texture on ammonium sorption has also been reported; e.g., Wang and Alva (2000) concluded that the sorption of NH_4^+ in sandy soils is much smaller than that in clay and silt soils. However, little is known about NH_4^+ sorption in wetland soils and the results of studies on NH_4^+ sorption in agricultural and forest soils do not necessarily fit NH_4^+ sorption in wetland soils. The desorption behaviors of NH_4^+ in wetland and terrestrial soils likely differ. Soil NH_4^+ levels in natural wetlands are greater than those in cultivated wetlands (Bedard-Haughn et al. 2006). Furthermore, the effect of wet land reclamation for farming on the NH_4^+ sorption behaviors has not been examined. The difference in NH_4^+ sorption and desorption between natural wetlands and farmlands could contribute to the difference in their NH_4^+ sink characteristics.

In the Sanjiang Plain of Northeast China, freshwater marshes have been converted to farmlands intensively, according to China's first national survey on wetland resources in 2003. This region experiences many freeze–thaw cycles (FTCs) in the winter of every year. Maehlum et al. (1995) reported that cold winter climate can affect hydraulic processes and biogeochemical processes of wetlands. Also, FTCs affect soil physical properties, microbial activity, and community composition (Lipson and Schmidt 2004; Six et al. 2004; Sjursen et al. 2005). FTCs result in the decreases in bulk density and penetration resistance in soils over the winter (Unger 1991). Oztas and Fayetorbay (2003) observed that in controlled laboratory

X. Yu, *Material Cycling of Wetland Soils Driven by Freeze–Thaw Effects*,
Springer Theses, DOI: 10.1007/978-3-642-34465-7_6,
© Springer-Verlag Berlin Heidelberg 2013

conditions, FTCs decrease soil aggregate stability, particularly in soils with high moisture content. Little is available on the sorption and desorption of NH_4^+ in the soils experiencing FTCs although many studies on NH_4^+ sorption potential of soils have been conducted under conditions with temperatures ranging from 5 to 25 °C (Lumbanraja and Evangelou 1994; Thompson and Blackmer 1992; Wang and Alva 2000). Wang et al. (2007) reported that the sorption and desorption of phosphorus in wetland soils is affected by FTCs and the freeze–thaw effect became more and more significant with the increase in FTC from one to six cycles. Currently, there has been growing interest in how changes in soil FTCs caused by climate change might alter soil nutrient dynamics (Henry 2007). The objective of this study was to investigate the effect of FTCs on the sorption and desorption of NH_4^+ in wetland soils and the potential impact of wetland reclamation on soil N pools.

6.1 Soil Properties

The differences in soil NH_4^+, K^+, and CEC (Table 6.1) were significant among the three sites DI-1 ($F = 9.773$, $P = 0.013$), DI-2 ($F = 5.413$, $P = 0.045$), and DI-3 ($F = 8.511$, $P = 0.017$) in Area DI. The same patterns were found only for CEC at Areas R ($F = 5.450$, $P = 0.045$) and DF ($F = 5.289$, $P = 0.047$). The soil K^+ concentration of Site R-3 was 54 % greater than the mean soil K^+ concentration of Sites R-1 and R-2. The soil Ca^{2+} concentration of Site R-3 was 23 % greater than the mean soil Ca^{2+} concentration of Sites R-1 and R-2. The soil K^+ concentration of Site DF-3 was 109 % greater than the mean soil K^+ concentration of Sites DF-1 and DF-2. The soil Ca^{2+} concentration of Site DF-3 was 24 % greater than the mean soil Ca^{2+} concentration of Sites DF-1 and DF-2. The mean soil NH_4^+ concentration of Sites DF-1 and DF-2 was 104 % greater than the soil NH_4^+ concentration of Site DF-3.

The differences in both soil CEC ($F = 5.257$, $P = 0.048$) and soil clay content ($F = 15.069$, $P = 0.004$) were significant among Sites R-1, DF-1, and DI-1. There was also significant difference in soil clay content ($F = 5.336$, $P = 0.046$) among Sites R-2, DF-2, and DI-2. The soil CEC of Site DI-2 was slightly greater than those of Sites R-2 and DF-2. In addition, the main clay mineral of soils in the three areas was hydrous mica, and the secondary clay minerals were vermiculite, montmorillonite, and kaolinite.

6.2 Sorption of NH_4^+ in Soils Subject to FTCs

Langmuir isotherm was selected to describe the sorption behaviors of NH_4^+ in soils because it provides an estimate for the sorption maximum of the adsorbent (Harter and Baker 1977; Vandenbruwane et al. 2007). Langmuir isotherm was expressed as:

$$C/S = k/S_{\mathrm{max}} + C/S_{\mathrm{max}}$$

Table 6.1 Soil physical and chemical properties (mean $\pm$ standard error) of the sites selected in different areas in the Sanjiang plain, Northeast China ($n = 3$)

Site	Organic matter (mg kg^{-1})	NH_4^+ (mg kg^{-1})	K^+ (g kg^{-1})	Ca^{2+} (g kg^{-1})	CEC (cmol kg^{-1})	Clay (%)	Soil textural
R-1	127.4 ± 24.3	33.5 ± 4.5	11.2 ± 4.1	69.7 ± 10.3	20.2 ± 5.5	4.2 ± 0.9	Sandy loam
R-2	72.1 ± 10.2	38.0 ± 3.9	10.6 ± 2.3	65.5 ± 24.8	41.6 ± 6.1	38.9 ± 3.3	Silt clay
R-3	115.8 ± 33.4	34.4 ± 3.4	16.8 ± 3.7	83.2 ± 20.5	14.3 ± 3.9	58.6 ± 16.5	Clay
DF-1	16.7 ± 1.9	104.0 ± 28.8	9.5 ± 3.9	59.8 ± 17.1	45.8 ± 7.7	50.8 ± 10.3	Clay
DF-2	29.3 ± 7.6	80.7 ± 10.7	10.0 ± 2.4	64.9 ± 13.5	39.0 ± 6.4	39.2 ± 7.7	Light clay
DF-3	68.5 ± 11.1	45.3 ± 5.5	20.4 ± 4.5	79.9 ± 23.2	13.6 ± 4.8	60.5 ± 6.2	Clay
DI-1	98.2 ± 18.8	76.1 ± 3.9	7.6 ± 1.1	77.3 ± 20.4	58.3 ± 5.5	78.4 ± 10.9	Heavy clay
DI-2	82.6 ± 5.7	71.9 ± 10.0	12.2 ± 3.2	62.0 ± 21.6	43.1 ± 8.6	70.1 ± 8.1	Clay
DI-3	114.1 ± 26.5	32.7 ± 5.2	22.5 ± 6.4	88.5 ± 29.1	16.4 ± 3.2	51.8 ± 7.2	Clay

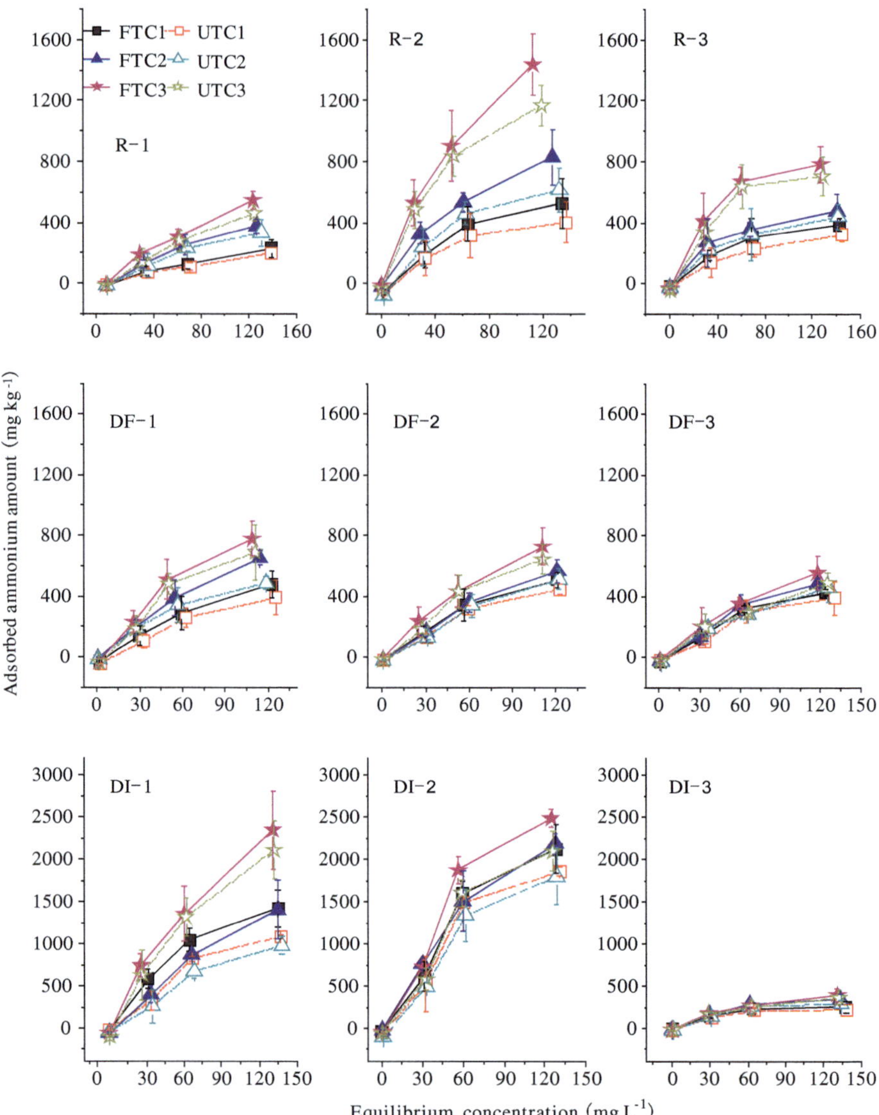

Fig. 6.1 Langmuir isotherms of NH$_4^+$ sorption in soils with and without freeze–thaw treatment. FTCs and UTCs represent the freeze–thaw cycles and control cycles. R-1, R-2, …, and DI-3 represent soils sampled at different sampling sites. Error bars represent the standard error of the mean of three replicated samples [Reprinted from Yu et al. (2011), Copyright (2011), with permission from Elsevier]

where S is the NH$_4^+$ sorption amount (mg kg^{-1}), C is the equilibrium concentration of NH$_4^+$ after equilibration (mg L^{-1}), S_{max} is the sorption maximum of NH$_4^+$ (mg kg^{-1}), used to estimate the sorption capacity, and k is a constant related to the bonding energy of NH$_4^+$ (L mg^{-1}).

Sorption isotherms of NH_4^+ in the FTT soils were always greater than those of the UT soils (Fig. 6.1). The nonparametric test indicated that the difference in the sorption amount of NH_4^+ between the FTT and UT soils was significant ($Z = -5.232$, $P < 0.01$). FTCs not only affected sorption amount of NH_4^+ in soils, but also influenced the maximum sorption (S_{max}) of NH_4^+. Values of S_{max} were significantly greater ($Z = -4.517$, $P < 0.01$) for the FTT soils than UT soils (Table 6.2). These results indicate that freeze–thaw could increase sorption capacity of NH_4^+. The sorption amount of NH_4^+ generally increased with the increase in FTCs with some exceptions; for example, the sorption amount of NH_4^+ in the

Table 6.2 Langmuir isotherms parameters (mean $\pm$ SE, $n = 3$) for NH_4^+ sorption in soils with freeze–thaw treatment (FTT) and without (UT). FTCs represent freeze–thaw cycles. S_{max} is maximum sorption capacity (mg kg^{-1}) used to estimate the sorption capacity. k is the Langmuir parameter related to the reciprocal of sorption energy

Site	FTT			UT		
	S_{max} (mg kg^{-1})	k (L kg^{-1})	R^2	S_{max} (mg kg^{-1})	k (mg kg^{-1})	R^2
FTC1						
R-1	815.4 ± 436.9	359.2 ± 252.7	0.99	704.9 ± 371.2	361.7 ± 249.7	0.99
R-2	1047.2 ± 494.5	126.1 ± 103.7	0.96	712.4 ± 366.0	98.0 ± 96.9	0.94
R-3	564.0 ± 85.4	63.9 ± 22.5	0.99	539.9 ± 132.4	96.9 ± 47.0	0.98
DF-1	1098.4 ± 288.1	139.1 ± 59.5	0.99	1052.4 ± 480.6	160.0 ± 114.4	0.98
DF-2	1625.0 ± 889.2	289.9 ± 212.2	0.99	1400.4 ± 1547.1	316.0 ± 460.2	0.96
DF-3	941.6 ± 448.8	144.9 ± 112.1	0.97	913.0 ± 614.5	165.8 ± 176.3	0.95
DI-1	2384.4 ± 298.4	89.7 ± 22.3	0.99	2327.7 ± 1045.1	148.7 ± 110.8	0.97
DI-2	4423.3 ± 2142.6	132.9 ± 108.4	0.96	3681.5 ± 2089.9	120.3 ± 119.7	0.94
DI-3	321.3 ± 45.4	32.8 ± 14.3	0.98	289.2 ± 56.3	32.5 ± 19.8	0.97
FTC3						
R-1	960.4 ± 250.1	194.9 ± 75.9	0.99	886.0 ± 329.9	210.2 ± 115.3	0.96
R-2	1559.7 ± 175.3	111.9 ± 22.5	1	1113.6 ± 502.5	101.3 ± 85.6	0.95
R-3	614.2 ± 69.2	43.5 ± 13.5	0.99	624.7 ± 94.5	61.0 ± 21.9	0.99
DF-1	1956.6 ± 332.0	226.7 ± 53.7	1	1318.7 ± 173.9	162.1 ± 33.1	1
DF-2	1706.4 ± 616.1	240.0 ± 120.9	0.99	1416.8 ± 856.2	226.3 ± 194.0	0.98
DF-3	1255.4 ± 655.9	184.0 ± 143.1	0.98	1167.5 ± 474.7	194.8 ± 117.9	0.99
DI-1	4679.8 ± 2033.2	310.8 ± 181.8	0.99	2854.6 ± 1790.1	256.2 ± 227.8	0.98
DI-2	4343.6 ± 747.5	123.7 ± 36.8	0.99	4264.7 ± 2748.5	168.3 ± 169.8	0.95
DI-3	482.4 ± 66.2	49.6 ± 17.1	0.99	411.4 ± 98.6	48.0 ± 29.5	0.97
FTC5						
R-1	1639.7 ± 534.8	248.7 ± 112.9	0.99	1491.5 ± 408.7	278.2 ± 103.4	1
R-2	2765.7 ± 151.2	104.2 ± 10.0	1	1845.4 ± 199.6	53.8 ± 15.3	1
R-3	1052.9 ± 137.6	39.6 ± 14.1	0.99	1008.1 ± 242.4	47.5 ± 28.9	0.97
DF-1	1830.1 ± 720.4	144.0 ± 88.4	0.98	1675.0 ± 1042.3	153.9 ± 147.1	0.96
DF-2	1876.0 ± 185.0	173.7 ± 25.5	1	1688.8 ± 723.4	175.6 ± 111.4	0.98
DF-3	1426.6 ± 289.7	182.4 ± 55.2	1	1220.1 ± 442.5	190.2 ± 103.5	0.99
DI-1	5348.1 ± 883.7	169.2 ± 44.2	1	4821.5 ± 1123.0	168.4 ± 62.2	0.99
DI-2	5050.9 ± 2395.1	122.5 ± 99.6	0.96	4629.1 ± 2833.1	145.0 ± 144.4	0.95
DI-3	620.5 ± 55.9	74.4 ± 14.2	1	556.9 ± 97.7	70.6 ± 26.8	0.99

R-1 soil increased by about 140 % after the 5th FTC (Fig. 6.1). These results indicated that FTCs increased the sorption capacity and sorption amount of NH_4^+, which confirmed the effect of FTCs on soil structure (Unger 1991). NH_4^+ can be fixed into the hexagonal holes on exposed surface between the sheets of 2:1 lattice-type clay minerals such as mica (Tang et al. 1996). FTCs can break down soil aggregates (Fitzhugh et al. 2001), and form new surfaces, which potentially increases the sorption of NH_4^+ to the exposed surfaces. We speculated that the FTCs increased the release of K^+, which provided more sorption sites for NH_4^+.

S_{max} of the wetland soil from Site DI (palustrine wetland) was 136–266 % greater than that from Site R (riverine wetland), and 146–181 % greater than that from Site DF (riverine wetland) (Table 6.2). This pattern might result from the greater clay content and CEC of the wetland soil from Site DI (Table 6.2). The river water washout might cause greater sand content and smaller clay content in the riverine wetland soils. The sorption capacity of NH_4^+ might be greater in the palustrine wetland soils than in the riverine wetland soils.

Comparing the sorption amounts of NH_4^+ among all the soils (Fig. 6.2), we found that the reclaimed farmland soils appeared to reduce the NH_4^+ sorption. For Areas DF and DI, the reclaimed farmland soils for soybean planting adsorbed less NH_4^+ than the wetland soils (Fig. 6.2). However, the values of S_{max} of the farmland soils were smaller than those of the wetland soils in each area (Table 6.2). These results indicate that the NH_4^+ sorption capacity in the farmland soils was weaker than that of the wetland soils. The K^+ and Ca^{2+} concentrations in the farmland soils were higher than those in the wetland soils (Table 6.2) and K^+ could reduce NH_4^+ sorption. Moreover, the altered hydroperiods could decrease CEC, which might be another reason for the reduced NH_4^+ sorption in the farmland soils. Phillips and Greenway (1998) observed that water logging increases CEC of soils with variably charged colloids. The increased CEC could result in the increased sorption of NH_4^+ through ion exchange. The shortened hydroperiod in the farmland might cause the decrease of CEC, and thus reduce sorption capacity of NH_4^+ in the farmland soils. Our results indicate that NH_4^+ concentration in the farmland soils was lower than that in the wetland soils at Areas DF and DI (Table 6.2), which is consistent with the results of Bedard-Haughn et al. (2006). Generally, farmland soils receive supplemental N through fertilizers.

We speculated that the reduced sorption capacity of NH_4^+ by reclamation might result in increasing loss of NH_4^+ from soils, where the amounts of N fertilizers were usually excessive. These additional N fertilizers were likely to pollute downstream waterbodies.

6.3 Desorption of NH_4^+ in the Soils Subject to FTCs

Desorption amounts of NH_4^+ decreased with the increase in number of desorptions. For both the FTT and UT soils, the desorption amount of NH_4^+ in the soils with 1,540 mg kg^{-1} NH_4^+ added in the sorption experiment was greater than that

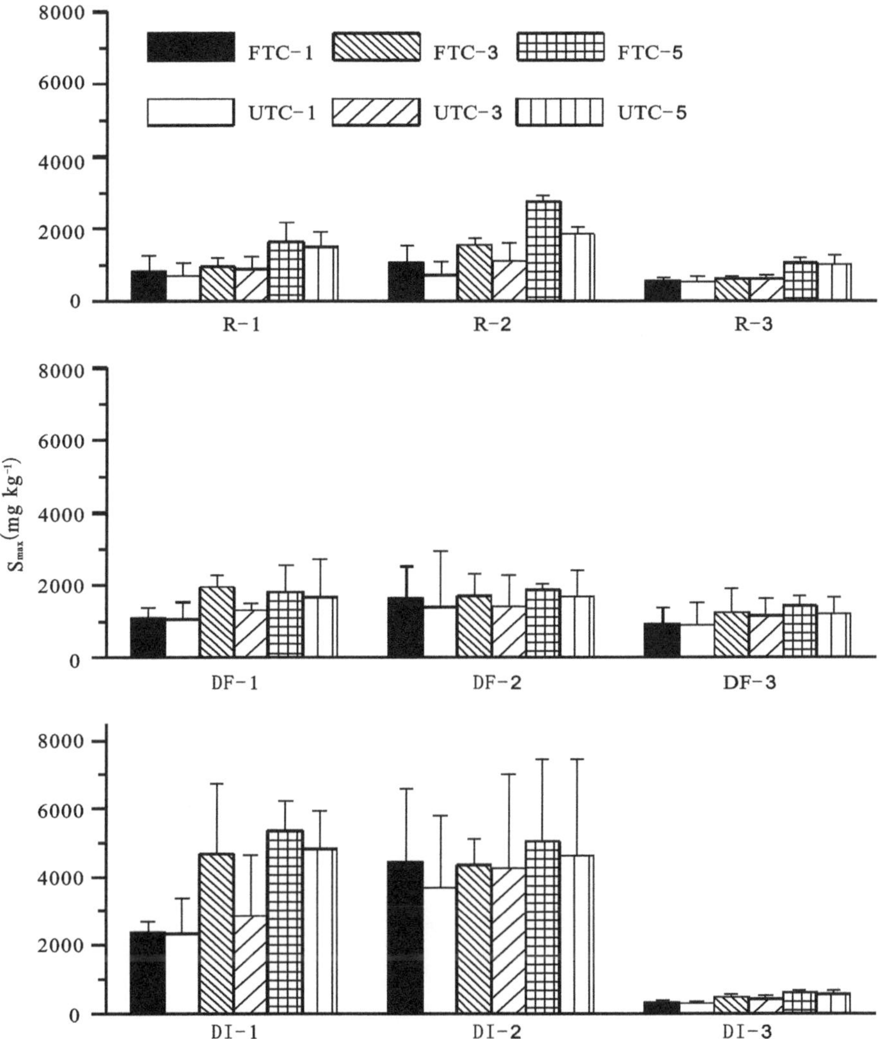

Fig. 6.2 The sorption maximum of NH$_4^+$ in varied soils with and without freeze–thaw treatment [Reprinted from Yu et al. (2011), Copyright (2011), with permission from Elsevier]

with 770 mg kg^{-1} NH$_4^+$ added. The desorption amount of NH$_4^+$ in the FTT soils was significantly ($P < 0.01$) less than that in the UT soils (Fig. 6.3). Moreover, the percentage of total desorption amount of NH$_4^+$ in desorption experiment in the amount of NH$_4^+$ adsorbed in the sorption experiment in the UT soils was significantly ($P < 0.01$) greater than that in the FTT soils (Table 6.3), which indicated that the desorption potential of NH$_4^+$ could be reduced by FTCs. The percentage of total desorption was the highest in the farmland soils of each area selected. For example, the percentage of total desorption in the R-3 soils (52–70 %) was greater than those in the R-1 and R-2 soils, which indicated that the NH$_4^+$ desorption

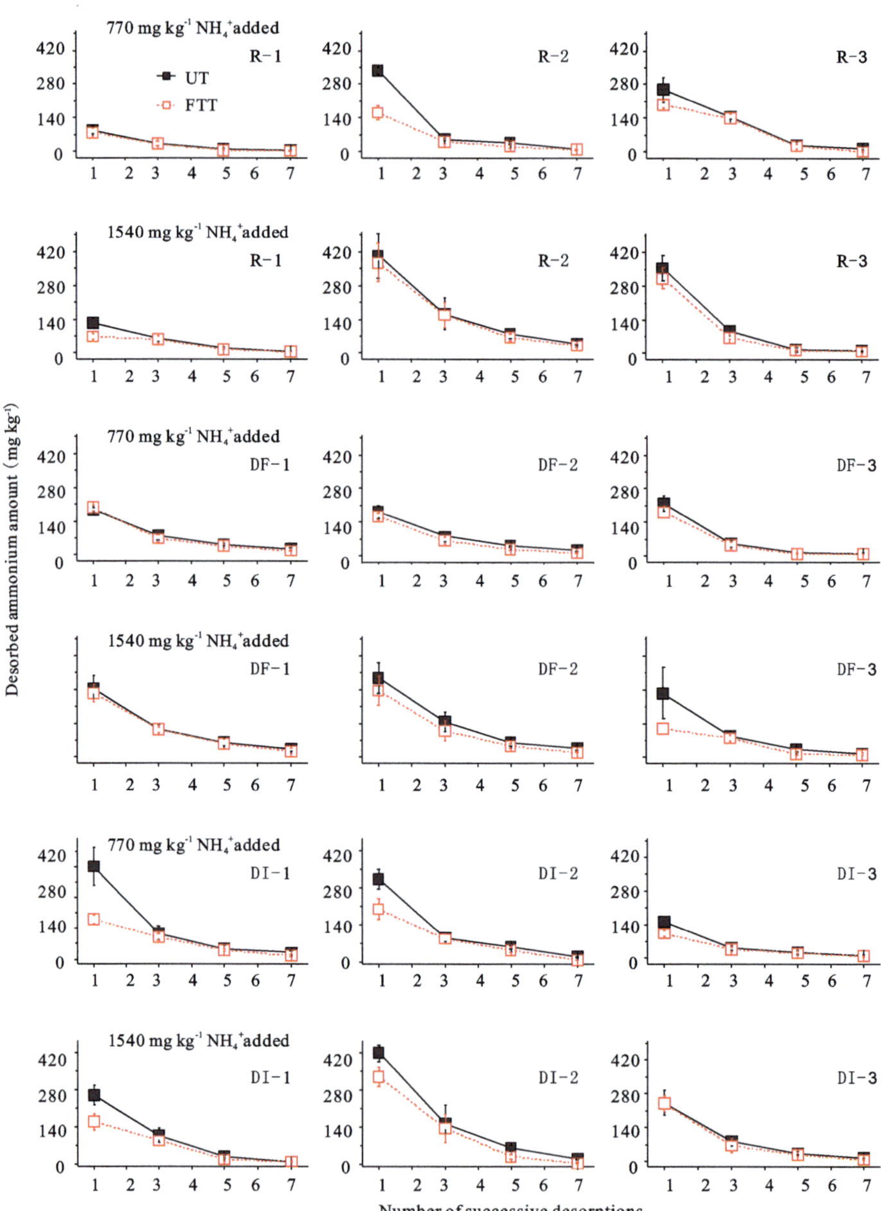

Fig. 6.3 Sequential desorption of NH_4^+ in soils with and without freeze–thaw treatment [Reprinted from Yu et al. (2011), Copyright (2011), with permission from Elsevier]

potential of the reclaimed farmland soils was stronger than that of the wetland soils. Moreover, the desorption potential in the DI-1 and DI-2 soils was greater than that in the R-1, R-2, DF-1, and DF-2 soils. These desorption potential patterns may be associated with the differences in clay content among these soils.

Table 6.3 The percentage (%) (Mean $\pm$ SE, $n = 3$) of total desorption amount of NH_4^+ in desorption experiment to the amount of NH_4^+ adsorbed in the sorption experiment

Site	FTT[a]	UT[a]	FTT[b]	UT[b]
R-1	39.3 $\pm$ 12.5	48.1 $\pm$ 17.6	25.9 $\pm$ 5.4	44.4 $\pm$ 11.4
R-2	25.7 $\pm$ 6.6	51.7 $\pm$ 7.1	43.5 $\pm$ 20.2	58.5 $\pm$ 13.7
R-3	53.5 $\pm$ 18.3	69.9 $\pm$ 26.5	51.8 $\pm$ 12.8	67.4 $\pm$ 13.6
DF-1	61.6 $\pm$ 27.4	69.8 $\pm$ 12.4	58.9 $\pm$ 19.5	71.8 $\pm$ 30.4
DF-2	61.1 $\pm$ 20.1	76.6 $\pm$ 29.3	62.0 $\pm$ 22.5	88.7 $\pm$ 32.2
DF-3	69.9 $\pm$ 28.1	96.8 $\pm$ 19.6	40.5 $\pm$ 15.3	82.1 $\pm$ 31.6
DI-1	25.0 $\pm$ 11.0	46.8 $\pm$ 17.2	25.6 $\pm$ 6.6	41.4 $\pm$ 19.4
DI-2	38.6 $\pm$ 15.7	63.5 $\pm$ 14.9	43.4 $\pm$ 11.9	66.8 $\pm$ 20.7
DI-3	64.5 $\pm$ 22.3	91.6 $\pm$ 29.1	84.6 $\pm$ 36.0	99.4 $\pm$ 35.8

[a] 770 mg kg^{-1} NH_4^+ added in sorption experiment
[b] 1540 mg kg^{-1} NH_4^+ added in sorption experiment

6.4 Ecological Implications of Freeze–Thaw Effect on NH_4^+

The increased sorption capacity and the decreased desorption potential of NH_4^+ caused by FTCs could increase NH_4^+ accumulation in the wetland soils. During winter when the activities of plants and microorganisms decrease, the NH_4^+ adsorbed may act as an intermediate N sink in soil (Matschonat and Matzner 1996). Thus, the increase of NH_4^+ retention in the wetland soils caused by FTCs was likely to increase N sink in the wetland soils. Furthermore, this increased N sink may increase N availability for plant uptake and reduce N leaching and NH_4^+ nitrification in soils after the freeze–thaw cycles. Due to the increase in sorption capacity of NH_4^+ caused by FTCs and anaerobic conditions in the wetlands, NO_2 emission, as a product of NH4 + NH_4^+ nitrification and denitrification, may be smaller in the wetland soils than in the farmland soils. Therefore, conversion of natural wetlands to crop production in the Sanjiang Plain might increase the release of NH_4^+ to downstream ecosystems.

6.5 Chapter Summary

Freeze–thaw significantly increased the sorption capacity of NH_4^+ and reduced the desorption potential of NH_4^+ in wetland soils. There were significant differences in the NH_4^+ sorption amount between soils with and without freeze–thaw treatment. The sorption amount of NH_4^+ increased with the increasing FTCs. The palustrine wetland soils had greater sorption capacity and weaker desorption potential of NH_4^+ than riverine wetland soils because of the greater clay content and CEC of the riverine soils. Because of the altered soil physical and chemical properties and hydroperiod, sorption capacity of NH_4^+ was smaller in farmland soils than in wetland soils, while desorption potential of farmland soils was higher than that of wetland soils. Thus, wetland reclamation will decrease sorption capacity and increase desorption potential of NH_4^+, which could result in N loss in farmland soils. Our results showed that FTCs might mitigate N loss from soils and reduce the risk of water pollution in downstream ecosystems.

References

Avnimelech Y, Laher M. Ammonium volatilization from soils: Equilibrium considerations [J]. Soil Sci Soc Am J. 1977;41:1080–4.

Bedard-Haughn A, Matson AL, Pennock DJ. Land use effects on gross nitrogen mineralization, nitrification, and N_2O emissions in ephemeral wetlands. Soil Biol Biochem. 2006; 38:3398–3406.

Fenn LB, Datcha JE, Wo E. Substitution of ammonium and potassium for added calcium in reduction of ammonia loss from surfaced-applied urea. Soil Sci Soc Am J. 1982;46:771–6.

Fitzhugh RD, Driscoll CT, Groffman PM, Tierney GL, Fahey TJ, Hardy JP. Effects of soil freezing, disturbance on soil solution nitrogen, phosphorus, and carbon chemistry in a northern hardwood ecosystem. Biogeochemistry. 2001;56:215–38.

Harter RD, Baker DE. Applications and misapplications of Langmuir equation to soil sorption phenomena. Soil Sci Soc Am J. 1977;41:1077–80.

Henry HAL. Soil freeze–thaw cycle experiments: trends, methodological weaknesses and suggested improvements. Soil Biol Biochem. 2007;39:977–86.

Lipson DA, Schmidt SK. Seasonal changes in an alpine soil bacterial community in the Colorado rocky mountains. Appl Environ Microbiol. 2004;70:2867–79.

Lumbanraja J, Evangelou VP. Adsorption–desorption of potassium and ammonium at low cation concentrations in three Kentucky subsoils. Soil Sci. 1994;157:269–78.

Maehlum T, Jenssen PD, Warne WS. Cold-climate constructed wetlands. Water Sci Technol. 1995;32:95–101.

Matschonat G, Matzner E. Soil chemical properties affecting NH_4^+ sorption in forest soils. J Plant Nutr Soil Sci. 1996;159:505–11.

Patrick WH, Reddy KR. Nitrification–denitrification reactions in flooded soils and water bottoms: dependence on oxygen supply and ammonium diffusion. J Environ Qual. 1976;5:469–72.

Phillips IR. Nitrogen availability and sorption under alternating waterlogged and drying conditions. Commun Soil Sci Plant Anal. 1999;30:1–20.

Phillips IR, Greenway M. Changes in water-soluble and exchangeable ions, cation exchange capacity, and phosphorus (max) in soils under alternating waterlogged and drying conditions. Commun Soil Sci Plant Anal. 1998;29:51–66.

Schadt CW, Martin AP, Lipson DA, et al. Seasonal dynamics of previously unknown fungal lineages in tundra soils. Science. 2003;301:1359–61.

Six J, Bossuyt H, Degryse S, et al. A history of research on the link between (micro) aggregates, soil biota, and soil organic matter dynamics. Soil Tillage Res. 2004;79:7–31.

Sjursen H, Michelsen A, Holmstrup M. Effects of freeze–thaw cycles on microarthropods and nutrient availability in a sub-Arctic soil. Appl Soil Ecol. 2005;28:79–93.

Tang Y, Feng K, Yin SX. Preferential fixation of ammonium to potassium by soils. Pedosphere. 1996;6:35–8.

Thompson TL, Blackmer AM. Quantity–Intensity relationships of soil ammonia in long-term rotation plots. Soil Sci Soc Am J. 1992;56:494–8.

Unger PW. Overwinter changes in physical-properties of no-tillage soil. Soil Sci Soc Am J. 1991;55:778–82.

Vandenbruwane J, De Neve S, Qualls RG, Sleutel S, Hofman G. Comparison of different isotherm models for dissolved organic carbon (DOC) and nitrogen (DON) sorption to mineral soil. Geoderma. 2007;139:144–53.

Wang FL, Alva AK. Ammonium adsorption and desorption in sandy soils. Soil Sci Soc Am J. 2000;64:1669–74.

Wang G, Liu J, Zhao H, Wang J, Yu J. Phosphorus sorption by freeze-thaw treated wetland soils derived from a winter-cold zone. Geoderma. 2007;138:153–61.

Yu XF, ZhangY X, Zou YC, et al. Sorption and desorption of ammonium in wetland soils subject to freeze-thaw. Pedosphere. 2011;21(2):251–8.

Chapter 7
Effect of Freeze–Thaw on the Mineralization of Organic Carbon, and Organic Nitrogen in Wetland Soil

The mineralization of organic carbon and organic nitrogen in soil is one of the key processes in the carbon and nitrogen cycles in wetland soil. In general it is believed that the mineralization of organic carbon and organic nitrogen in soil mainly depends on the moisture and temperature. Conditions of soil moisture affect the soil respiration rate (Mainiero and Kazda 2004) by restricting the penetration of oxygen and the types of microorganisms, but temperature affects the mineralization rates of organic carbon and organic nitrogen in soil by affecting microbial activity (Reichstein et al. 2000). Because of the interaction between temperature and moisture, the real mineralization rate in a soil often depends on both moisture and temperature conditions at the same time.

Freeze–thaw effects can lead to the death of some microorganisms (Lipson and Monson 1998; Fitzhugh et al. 2001), and the consequent reduction in the microorganism population will reduce CO_2 emissions and the concentration of inorganic nitrogen, thus reducing the amount of mineralized organic carbon and organic nitrogen in a soil (Vestgarden and Austnes 2009). Freeze–thaw can, however, promote the death of some soil microorganisms, and the death of these microorganisms will release a certain amount of CO_2 and inorganic nitrogen, and the broken dead cells will also release nutrients which will aid in the survival of other live microorganisms, which will in turn increase the respiration activities of the surviving microorganisms. Thus, freeze–thaw effects on the actual mineralization of organic carbon and organic nitrogen in a soil depend on the balance between: (1) the reduced amounts of CO_2 and inorganic nitrogen caused by the reduction in the microbial community, and (2) the increment of CO_2 and inorganic nitrogen released from dead microbes resulting from freezing and thawing.

Through the simulation experiments described in this chapter, the mineralization processes of organic carbon and organic nitrogen in different types of wetland soils were studied under freeze–thaw conditions, and the amount of organic carbon and organic nitrogen mineralization was calculated. The relationship between freeze–thaw effects and the mineralization of organic carbon and organic nitrogen in the wetland was then investigated.

X. Yu, *Material Cycling of Wetland Soils Driven by Freeze–Thaw Effects*,
Springer Theses, DOI: 10.1007/978-3-642-34465-7_7,
© Springer-Verlag Berlin Heidelberg 2013

7.1 Effect Freeze–Thaw on the Mineralization of Organic Carbon in the Soil of Wetland

7.1.1 Soil Physicochemical Properties

Physicochemical properties of the sampled soils are shown in Table 7.1.

7.1.2 Freeze–Thaw Effects on the Mineralization of Organic Carbon in the Wetland Soil

The dynamics of CO_2–C released from mineralization of the *Carex lasiocarpa* wetland soil, the *Calamagrostis angustifolia* wetland soil and the dryland soil with freeze–thaw treatment or the control group during the 26-day culture period are shown in Fig. 7.1. It shows that changes in the amounts of mineralization for the *Carex lasiocarpa* wetland soil, the *Calamagrostis angustifolia* wetland soil and the dryland soil are basically the same for both the freeze–thaw treated group and the control group, and the fast-slow turning point of soil CO_2–C release dynamics appears between the 7th and 10th days. However, the mineralization amount of organic carbon in the soil after freeze–thaw treatment is higher than that of the control, and the mineralization amount of organic carbon in the *Carex lasiocarpa* wetland soil is higher than that in the *Calamagrostis angustifolia* wetland and dryland soils, and the difference of organic carbon mineralization amount between the *Calamagrostis angustifolia* wetland and dryland soils is small.

Amounts of mineralization of organic carbon during the culture period were analyzed by analysis of variance (ANOVA) (Table 7.2). The freeze–thaw treatment, soil type and incubation time as well as their pair-wise interactions all have significant impacts on the mineralization amount of organic carbon in soil.

A first-order kinetic equation was used to fit the soils organic carbon mineralization amount of the *Carex lasiocarpa* wetland, the *Calamagrostis angustifolia* wetland and the dryland under the freeze–thaw treatment and the control experimental conditions.

$$C_t = C_0 \left(1 - e^{-kt}\right)$$

Table 7.1 Selected physical and chemical properties of soil samples ($n = 3$)

	H	M	G
Clay (%)	40.2 ± 11.2	64.5 ± 8.09	46.8 ± 4.12
pH	5.6 ± 0.5	5.4 ± 0.6	6.0 ± 0.5
OM (%)	12.9 ± 3.13	11.7 ± 3.27	8.3 ± 3.09
DOC (mg kg^{-1})	2762.00 ± 182.00	546.00 ± 76.20	197.10 ± 31.50
NH_4^+-N (mg kg^{-1})	648.51 ± 102.31	871.74 ± 79.69	125.36 ± 36.97
NO_3^--N (mg kg^{-1})	2.25 ± 0.33	2.10 ± 0.45	8.32 ± 2.14

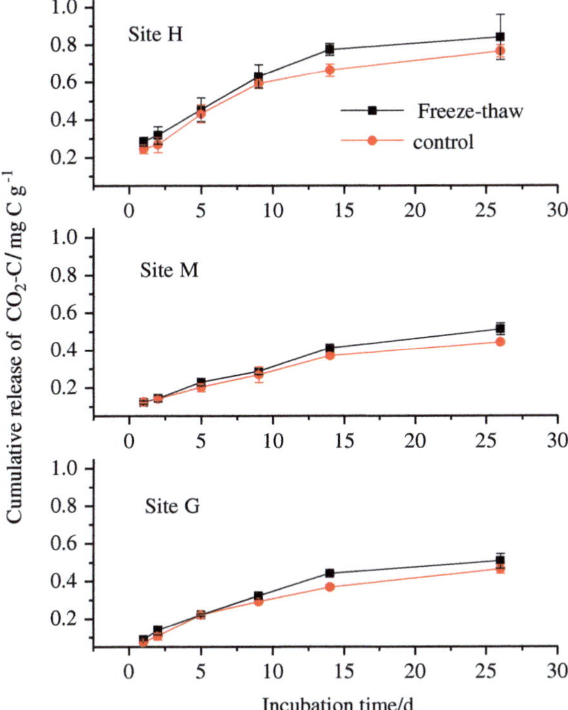

Fig. 7.1 Dynamics of accumulative amounts of C mineralized with incubation days in soils with freeze–thaw treatment

Table 7.2 Results of ANOVA for accumulative amounts of C mineralized in soils with freeze–thaw treatment

Influence factor	F	P
Soil type	1910.790	0.000
Incubation time	1363.805	0.000
Freeze–thaw	109.444	0.000
Soil type × incubation time	33.800	0.000
Soil type × freeze–thaw	5.386	0.026
Freeze–thaw × incubation time	7.021	0.005

Where Ct (mg C g^{-1}) is the amount of mineralization at incubation time t (d), C_o (mg C g^{-1}) is the potential mineralization amount of soil organic carbon, k (d^{-1}) is the mineralization rate constant of soil organic carbon, and t (d) is the incubation time.

Table 7.3 shows that, during the incubation period, a first-order kinetic equation describes well the dynamics of soil organic carbon mineralization under the freeze–thaw treatment and the control conditions ($R^2 > 0.897$). For the freeze–thaw treatment, the potential mineralization amounts of organic carbon in the

Table 7.3 First-order kinetic parameters and C_0/SOC in soils with freeze–thaw treatment

Site	Treatment	C_0 (mg C g^{-1})	k (d^{-1})	R^2	C_0/SOC (%)
H	Freeze–thaw	0.812	0.203	0.897	0.87
	Control	0.727	0.207	0.929	0.78
M	Freeze–thaw	0.529	0.110	0.923	0.62
	Control	0.441	0.130	0.904	0.52
G	Freeze–thaw	0.533	0.116	0.982	0.89
	Control	0.478	0.112	0.989	0.80

Carex lasiocarpa wetland soil, the *Calamagrostis angustifolia* wetland soil and the dryland soil are 0.812, 0.529, and 0.533 mg C g^{-1}, respectively, and 0.727, 0.441, and 0.478 mg C g^{-1}, respectively, for the control group. The potential mineralization amount of organic carbon after the freeze–thaw treatment is larger than that in the control for the *Carex lasiocarpa* wetland, the *Calamagrostis angustifolia* wetland and the dryland soil, which indicates that freeze–thaw increased the potential mineralization of soil organic carbon.

The potential mineralization amount of organic carbon in the *Carex lasiocarpa* wetland soil is greater than that in the *Calamagrostis angustifolia* wetland soil and dryland soil, while the potential mineralization amount of organic carbon in the dryland soil is slightly greater than that in the *Calamagrostis angustifolia* wetland soil. However, the C_0/SOC value of the dryland soil is larger than for both types of wetland soil, and the C_0/SOC value of the *Carex lasiocarpa* wetland soil is larger than that of the *Calamagrostis angustifolia* wetland soil. Therefore, although the potential mineralization amount of organic carbon in the dryland soil is close to that in the *Calamagrostis angustifolia* wetland soil, the extent of mineralization of organic carbon in the dryland soil was greater than that in both types of wetlands soil within the 26 days of incubation.

The mineralization rate of total organic carbon for the *Carex lasiocarpa* wetland soil, the *Calamagrostis angustifolia* wetland soil and the dryland soil in the freeze–thaw treated group and control group during the whole incubation period is shown in Fig. 7.2. The mineralization rate of soil organic carbon is the percentage of mineralized organic carbon over the total organic carbon in the soil during a period of time, which reflects the difficulty level of soil organic carbon mineralization. During the entire 26-day incubation, with the freeze–thaw treatment, the mineralization rates of total organic carbon were 0.90, 0.61 and 0.84 % for the *Carex lasiocarpa* wetland soil, the *Calamagrostis angustifolia* wetland soil and the dryland soil, respectively, whereas in the control conditions, the mineralization rates of total organic carbon were 0.82, 0.52 and 0.77 %, respectively, for these three kinds of soil.

It can be seen that the mineralization rate of soil total organic carbon was increased after the freeze–thaw treatment. The mineralization rate of total organic carbon in the *Carex lasiocarpa* wetland soil was greater than that in the *Calamagrostis angustifolia*

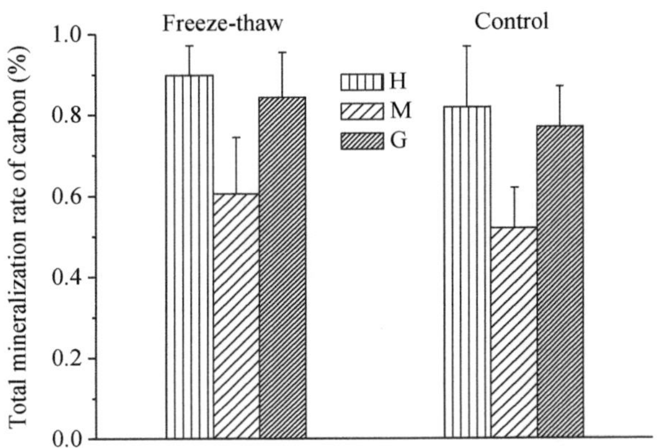

Fig. 7.2 Total rates of C mineralized in soils with freeze–thaw treatment

wetland soil, and that in the dryland soil, which was reclaimed from *Calamagrostis angustifolia* wetland, was significantly higher than that in the *Calamagrostis angustifolia* wetland soil. ANOVA was performed on the accumulated mineralization rate of total organic carbon during the incubation period. It indicated that both the freeze–thaw treatment ($F = 72.703$; $p < 0.0001$) and soil type ($F = 334.447$; $p < 0.0001$) significantly affected the cumulative mineralization rate of soil organic carbon, but their interaction on the mineralization rate of soil organic carbon was not significant ($F = 0.101$; $p = 0.905$).

It can be seen from the above findings that, during the 26-day incubation period, the mineralization amount and the potential mineralization amount of organic carbon, C_0/SOC values and mineralization rate of soil total organic carbon with the freeze–thaw treatment are all larger than those under the control conditions for the *Carex lasiocarpa* wetland, the *Calamagrostis angustifolia* wetland and the dryland. This indicates that freeze–thaw can increase the mineralization amount of soil organic carbon, increase the degree of mineralization of the soil organic carbon and change the difficulty with which the soil organic carbon is mineralized. The freeze–thaw effect on the mineralization of soil organic carbon appears to mainly be due to its impact on the amount, composition and activity of the soil microorganisms.

When the freezing temperature is below -5 °C, freeze–thaw can lead to the death of some microorganisms, which will affect the mineralization of soil organic carbon (Lipson and Monson 1998; Fitzhugh et al. 2001). Under the freeze–thaw conditions, CO_2 released from the mineralization of soil organic carbon mainly comes from: CO_2 released from the death of microbes, caused by the freeze–thaw process, and CO_2 generated by the respiration of surviving microorganisms. In a study of soil organic carbon mineralization in an indoor cultivated soil, Schimel and Clein (1996) reported that freeze–thaw reduced soil CO_2 emission which is the opposite of the results in this study. However, Vestgarden and Austnes (2009) have recently shown that the freeze–thaw process can increase the mineralization of soil organic carbon.

It is known that the freeze–thaw process can lead to the death of some soil microorganisms. The death of these microorganisms releases a certain amount of CO_2, and the broken dead cells also release nutrients for the surviving microorganisms to use, thus increasing to some extent the respiratory activity of the surviving microorganisms. In the meantime, the reduced numbers of microorganisms will also cause reduced CO_2 emission. Effects of the freeze–thaw process on the mineralization of soil organic carbon should come from the positive and negative offset of the released CO_2 from the freeze–thaw-induced death of microorganisms and reduced respiration from the reduction in microbial biomass.

Since soil microorganisms can tolerate a decrease in temperature to a certain extent, if the designed freezing temperature is not low enough in indoor incubation experiments, the observed effects of freeze–thaw on the mineralization of soil organic carbon will not be significant. This might also be a reason for the inconsistent results between this study and that done by Schimel and Clein (1996). In this study, the freezing temperature of the freeze–thaw treatment was -15 °C, but it was -5 °C in Schimel and Clein's (1996) study. In their conditions with a freezing temperature of -5 °C, only a limited number of microorganisms will have died, with a lesser reduction in the microorganism population, and lower respiratory activity at the low temperature. These may all have contributed to the reduced mineralization of organic carbon.

In addition, different types of soil have big differences in their tolerance to freeze–thaw, and therefore the effect of freeze–thaw on the mineralization of organic carbon in different types of soil are often completely different under the same experimental conditions (Neilsen et al. 2001). Furthermore, because of global warming, the freeze–thaw period in the temperate regions will be extended, which would probably increase the CO_2 emission amounts in wetland soils and further form a vicious circle. Meanwhile, if global warming increases the frozen temperature to slightly below 0 °C, the amounts of CO_2 emitted from wetland soils could probably be reduced instead. Therefore, the response of organic carbon mineralization in wetland soils under freeze–thaw conditions to climate warming requires further in-depth study.

7.2 The Effect of Freeze–Thaw on the Mineralization of Organic Nitrogen in Wetland Soil

ANOVA was performed on the mineralization amount of organic nitrogen during the incubation period (Table 7.4). The analysis showed that the freeze–thaw treatment, soil type and incubation time as well as their pair-wise interaction all have a significant impact on the mineralization amount of soil organic nitrogen.

Dynamics of organic nitrogen mineralization amount of the *Carex lasiocarpa* wetland soil, the *Calamagrostis angustifolia* wetland soil and the dryland soil with the freeze–thaw treatment and the control conditions are shown in Fig. 7.3.

Table 7.4 Results of ANOVA for accumulative amounts of N mineralized in soils with freeze–thaw treatment

Influence factor	F	P
Soil type	774.567	0.000
Incubation time	407.808	0.000
Freeze–thaw	267.203	0.000
Soil type × incubation time	1705.789	0.000
Soil type × freeze–thaw	1560.893	0.000
Freeze–thaw × incubation time	4.176	0.026

Fig. 7.3 Dynamics of accumulative amounts of N mineralized with incubation days in wetland soils with freeze–thaw treatment

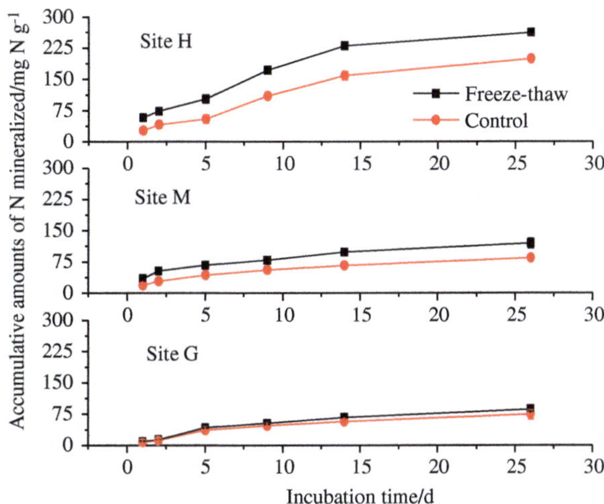

During the 26-day incubation period, with the freeze–thaw treatment, mineralization amounts of organic nitrogen in the *Carex lasiocarpa* wetland soil, the *Calamagrostis angustifolia* wetland soil and the dryland soil were, respectively, 1.32–2.25 times, 1.44–1.99 times and 1.17–1.41 times that of the corresponding soil in the control conditions, which suggests that the freeze–thaw process can increase the mineralization amount of soil organic nitrogen. One reason is that part of the soil microbial biomass is killed by the freezing and thawing, which results in cell rupture and the release of some mineral nitrogen. The other reason is that the dead microorganisms provide a matrix for the residual microbial biomass, which stimulates their activity and also promotes the mineralization of soil organic nitrogen. Additionally, freeze–thaw can lead to the fragmentation of soil aggregates and soil colloids, thus releasing some NH_4^+–N.

Figure 7.3 shows that the mineralization amount of organic nitrogen in the *Carex lasiocarpa* wetland soil is greater than that in the *Calamagrostis angustifolia* wetland soil or that in the dryland soil. Correlation analysis was performed on the mineralization amount of soil total organic nitrogen within the 26-day incubation period and soil physicochemical properties. It showed a significant correlation

($r = 0.998$; $p = 0.035$) between the content of soil DOC and the mineralization amount of soil total organic nitrogen. However, the correlation was not significant between the mineralization amounts of soil total organic nitrogen and soil organic matter, or NH_4^+–N, or NO_3^-–N content. This is because soil DOC is an intermediate in the conversion of soil organic matter and the microbial metabolic activity, and also the soil DOC content is a reflection of the decomposition and utilization of organics by the microbial biomass.

The mineralization amount of soil organic nitrogen for the *Carex lasiocarpa* wetland, the *Calamagrostis angustifolia* wetland and the dryland under the freeze–thaw treatment conditions and the control conditions was fitted with a first-order kinetic equation:

$$N_t = N_0 \left(1 - e^{-kt}\right)$$

where, N_t (mg N g^{-1}) is amount of mineralization at the incubation time t (d), N_0 (mg N g^{-1}) is the potential mineralization amount of soil organic nitrogen, k (d^{-1}) is the mineralization rate constant of soil organic nitrogen, and t (d) is the incubation time.

Values of all parameters in the equation are shown in Table 7.5. The changes in mineralization amount of soil organic nitrogen over time ($R^2 > 0.837$) under the freeze–thaw and control conditions can be well described by the first-order kinetic equation for the *Carex lasiocarpa* wetland, the *Calamagrostis angustifolia* wetland and the dryland. ANOVA analysis was performed on the potential mineralization amount of soil organic nitrogen in the *Carex lasiocarpa* wetland, the *Calamagrostis angustifolia* wetland and the dryland under the freeze–thaw and control conditions. It showed that freeze–thaw significantly affected the potential mineralization amount of soil organic nitrogen. The potential mineralization amount of soil organic nitrogen under the freeze–thaw treatment was higher than that of soil under the control conditions.

The dynamics of the mineralization amount of organic nitrogen versus time under the freeze–thaw treatment conditions during the incubation period are shown in Fig. 7.4. The ratio of the mineralization amount of soil organic nitrogen under the freeze–thaw treatment conditions over that under the control conditions reflects the impact level of freeze–thaw on the mineralization amount of soil organic nitrogen. Figure 7.4 shows that the ratio is the highest in the first

Table 7.5 First-order kinetic parameters in soils with freeze–thaw treatment

Site	Treatment	$N_0 \left(\text{mg N} g^{-1}\right)$	$k \left(d^{-1}\right)$	R^2
H	Freeze–thaw	276.520	0.116	0.958
	Control	243.417	0.067	0.978
M	Freeze–thaw	107.085	0.229	0.837
	Control	80.060	0.148	0.950
G	Freeze–thaw	91.658	0.100	0.988
	Control	78.202	0.098	0.988

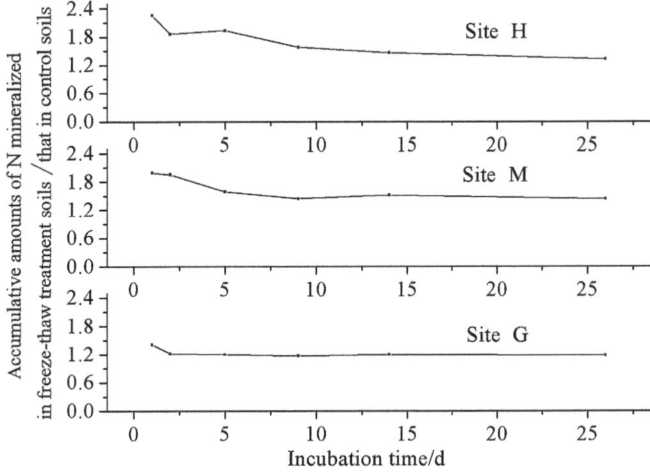

Fig. 7.4 Ration of accumulative amounts of N mineralized in freeze–thaw treatment soils to that in control soils with incubation days

freeze–thaw cycle and then overall it has a downward trend, which suggests that the impact of the freeze–thaw process on the mineralization amount of soil organic nitrogen decreases gradually with an increase in the number of freeze–thaw cycles. Additionally, the freeze–thaw process has a larger influence on the mineralization amount of soil organic nitrogen in the *Carex lasiocarpa* wetland than in the *Calamagrostis angustifolia* wetland and the dry land.

7.3 Chapter Summary

In this chapter, the mineralization of soil organic carbon and soil organic nitrogen under the freeze–thaw conditions were studied through simulation experiments with different soil types from the wetland. We showed that the freeze–thaw conditions could significantly increase the mineralization amount of wetland soil organic carbon, enhance the mineralization of soil organic carbon, and change the difficulty level of soil organic carbon mineralization. The mineralization amount of soil organic carbon in the *Carex lasiocarpa* wetland was higher than that in both the *Calamagrostis angustifolia* wetland and the dryland, and the mineralization level of the dryland soil organic carbon was higher than that in the *Carex lasiocarpa* wetland soil and in the *Calamagrostis angustifolia* wetland soil. Freeze–thaw can significantly increase the mineralization amount of organic nitrogen in wetland soil, and it has a larger impact on the mineralization amount of soil organic nitrogen in the *Carex lasiocarpa* wetland than in the *Calamagrostis angustifolia* wetland and the dryland. The mineralization amount of soil organic nitrogen in the *Carex lasiocarpa* wetland was larger than in the *Calamagrostis angustifolia* wetland and the dry land, and also significantly correlated with the DOC content. Additionally, because of global warming, the freeze–thaw period in the temperate region is likely to be extended, which would probably increase amounts of CO_2 emitted in the wetland soil and further form a vicious circle. Meanwhile, if global warming causes the freezing temperature to increase to slightly below 0 °C, the amount of CO_2 emitted by the wetland soil could probably be reduced. Therefore, the response of organic carbon mineralization in the wetland soils under the freeze–thaw process to climate warming requires further in-depth study.

References

Fitzhugh RD, Driscoll CT, Groffman PM, Tierney GL, Fahey TJ, Hardy JP. Effects of soil freezing, disturbance on soil solution nitrogen, phosphorus, and carbon chemistry in a northern hardwood ecosystem. Biogeochemistry. 2001;56:215–38.

Lipson DA, Monson RK. Plant–microbe competition for soil amino acids in the alpine tundra: effects of freeze–thaw and dry–rewet events. Oecologia. 1998;113:406–14.

Mainiero R, Kazda M. Effects of Carex rostrata on soil oxygen in relation to soil moisture. Plant Soil. 2004;270:311–20.

Neilsen CB, Groffman PM, Hamburg SP, et al. Freezing effects on carbon and nitrogen cycling in northern hardwood forest soils. Soil Sci Soc Am J. 2001;65:1723–30.

Reichstein M, Bednorz F, Broll G, Kätterer T. Temperature dependence of carbon mineralisation: conclusions from along–term incubation of subalpine soil samples. Soil Biol Biochem. 2000;32:947–58.

Schimel JP, Clein JS. Microbial response to freeze–thaw cycles in tundra and taiga soils. Soil Biol Biochem. 1996;28:1061–6.

Vestgarden LS, Austnes K. Effects of freeze–thaw on C and N release from soils below different vegetation in a montane system: a laboratory experiment. Glob Change Biol. 2009;15:876–87.

Chapter 8
Response of Soil Constituents to Freeze–Thaw Cycles in Wetland Soil Solution

Freezing and thawing of soils is a common phenomenon in the winter-cold zone, which can subsequently affect carbon, nitrogen and phosphorus cycling in soils and the leaching of nutrients through influencing biochemical and physicochemical processes. The soil solution nutrient pool in the freeze–thaw period may control the quantity of nutrients available to plants in the following spring (Buckeridge and Grogan 2008).

Little is known about the effects of freeze–thaw cycles on wetland soils. In spring, wetlands in the winter-cold zone are usually flooded by snow and ice melt, so that wetland soils become supersaturated with water. Under the flooded hydrological condition, wetlands are more frequently coupled to adjacent ecosystems through dissolved chemical exchanges. The freeze–thaw effects on dissolved carbon, nitrogen and phosphorus in wetland soil solutions should have close relation to the dissolved carbon, nitrogen and phosphorus losses and fluxes from wetland soils in the freeze–thaw season. In recent years, more and more wetlands have been reclaimed for farming in China, and nitrogen and phosphorus fertilizers are applied to the soils. The release of dissolved nitrogen and phosphorus from farmland soils subjected to freeze–thaw cycles is related to the downward wetland water quality, because there is little plant uptake during freeze–thaw season.

Contrasting results concerning the effects of freeze–thaw cycles on nutrient cycling have been found in some studies with forest, grassland and tundra soils. Increased soil emissions of CO_2 and N_2O (Prieme and Christensen 2001; Muller et al. 2002; Ludwig et al. 2006) and increases of dissolved organic carbon (DOC) and NH_4^+–N in soils after freeze–thaw cycles (Schimel and Clein 1996; Herrmann and Witter 2002; Zhao et al. 2010) have been reported. The increased leaching of NO_3^-–N from soils contributed to the high NO_3^-–N concentrations in the water bodies after freeze–thaw cycles (Hentschel et al. 2008). At the same time, Wang et al. (2007) observed an increase of phosphorus sorption and a decrease of phosphorus desorption in wetland soils after freeze–thaw cycles. The release from microbial biomass (Ivarson and Sowden 1970; DeLuca et al. 1992; Matzner and Borken 2008), root turnover (Tierney et al. 2001) and change in soil

X. Yu, *Material Cycling of Wetland Soils Driven by Freeze–Thaw Effects*,
Springer Theses, DOI: 10.1007/978-3-642-34465-7_8,
© Springer-Verlag Berlin Heidelberg 2013

structure (van Bochove et al. 2000) caused by freeze–thaw cycles were the main reasons for the release of carbon and nitrogen from soils during freeze–thaw season. Freezing temperature and freeze–thaw frequency might influence the freeze–thaw effects (Matzner and Borken 2008). Furthermore, the disturbance of soil samples was a major drawback in many freeze–thaw studies (Henry 2007). Some freeze–thaw studies have focused on carbon and nitrogen release from two or three soil depths of undisturbed soil columns (Teepe et al. 2001; Goldberg et al. 2008; Hentschel et al. 2008). An investigation of DOC, NH_4^+–N, NO_3^-–N and the total dissolved phosphorus (TDP) in soil columns with a higher resolution in depth might be interesting and important.

Chemical changes in soil solutions over time could reflect physicochemical processes such as dissolved matter transfer and transport (Zou et al. 2009). The concentrations of soil solution contents can also indicate the fluxes of dissolved nutrients, with the importance to the soil nutrient losses and water quality. In previous research of freeze–thaw effects on carbon and nitrogen concentrations in soil solutions (Lipson and Monson 1998; Grogan et al. 2004), the soil solutions were mainly extracted by K_2SO_4 solution, but the process could destroy soil samples and disturb the chemical composition and the equilibrium of soil solutions. Using a soil solution sampler, soil solution can be extracted from specific soil layers without disturbing its chemical composition.

The objectives of this study were to study the freeze–thaw effects on DOC, NH_4^+–N, NO_3^-–N and TDP concentrations in wetland and farmland soil solutions, and to investigate the spatial heterogeneity of the freeze–thaw effects.

8.1 DOC in Soil Solutions

The freeze–thaw treatment, cycle, soil layer and sampling site had significant effects on DOC concentrations in soil solutions (Table 8.1, 8.2). The DOC concentrations in soil solutions were not significantly affected by freeze–thaw treatment × cycle interaction and freeze–thaw treatment × soil layer interaction (Table 8.1, 8.2), but were significantly affected by freeze–thaw treatment × site interaction (Table 8.2, Fig. 8.1). In 111 of 112 paired data (freeze–thaw treatment and control), DOC concentrations in soil solutions increased after freeze–thaw treatment (Fig. 8.2). In 12 of 16 soil layers of treatment columns, we observed an increase of DOC concentration in the early two freeze–thaw cycles (Fig. 8.2).

The DOC increments caused by freeze–thaw treatment varying with cycles showed no consistent trend among various soil types or layers (Fig. 8.2). Therefore, we calculated the average increments of seven cycles, and compared the differences of DOC increments in various sampling sites and soil layers (Fig. 8.3). The average DOC increments were highly different between sampling sites in the order of *Carex* marsh > *Carex* marshy meadow > *Calamagrostis* wet grassland > soybean field in each soil layer, except for *Carex* marsh < *Carex* marshy meadow in the 30–40 cm soil layer. The average DOC increments in the upper three soil layers (0–10, 10–20

Table 8.1 MANOVA results (p values) for DOC, NH_4^+–N, NO_3^-–N and TDP concentration in soil solution with two treatments, seven cycles and four soil layers ($n = 3$). Boldface values indicate significant effects at $p < 0.05$

Variables		Freeze–thaw treatment	Cycle	Soil layer	Freeze–thaw treatment × cycle	Freeze–thaw treatment × soil layer	Cycle × soil layer	Freeze–thaw treatment × cycle × soil layer
DOC	*Carex* marsh	<0.0001	<0.0001	0.006	0.181	0.06	0.125	0.552
	Carex marshy meadow	<0.0001	<0.0001	<0.0001	0.161	0.082	0.232	0.708
	Calamagrostis wet grassland	<0.0001	0.02	<0.0001	0.397	0.165	0.261	0.924
	Soybean field	<0.0001	<0.0001	<0.0001	0.429	0.457	0.033	0.602
NH_4^+–N	*Carex* marsh	<0.0001	<0.0001	<0.0001	0.282	0.239	<0.0001	0.433
	Carex marshy meadow	<0.0001	<0.0001	<0.0001	0.634	0.056	0.071	0.996
	Calamagrostis wet grassland	<0.0001	<0.0001	0.03	0.056	0.428	0.001	0.460
	Soybean field	<0.0001	<0.0001	<0.0001	0.513	<0.0001	0.001	0.322
NO_3^-–N	*Carex* marsh	<0.0001	<0.0001	<0.0001	0.04	0.29	<0.0001	0.578
	Carex marshy meadow	<0.0001	0.001	<0.0001	0.045	0.422	0.001	0.709
	Calamagrostis wet grassland	<0.0001	<0.0001	<0.0001	0.001	0.054	<0.0001	0.261
	Soybean field	<0.0001	<0.0001	<0.0001	0.035	0.414	<0.0001	0.630
TDP	*Carex* marsh	<0.0001	<0.0001	<0.0001	0.08	0.065	0.043	0.390
	Carex marshy meadow	<0.0001	<0.0001	<0.0001	0.189	0.058	0.318	0.531
	Calamagrostis wet grassland	<0.0001	0.008	<0.0001	0.546	0.156	0.349	0.894
	Soybean field	<0.0001	<0.0001	<0.0001	0.109	0.178	0.002	0.334

Table 8.2 MANOVA results (p values) for DOC, NH_4^+–N, NO_3^-–N and TDP concentration in soil solution with two treatments, seven cycles and four sampling sites ($n = 3$). Boldface values indicate significant effects at $p < 0.05$

	Variables (cm)	Freeze–thaw treatment	Cycle	Site	Freeze–thaw treatment × cycle	Freeze–thaw treatment × site	Cycle × site	Freeze–thaw treatment × cycle × site
DOC	0–10	<0.0001	<0.0001	<0.0001	0.058	<0.0001	0.05	0.633
	10–20	<0.0001	<0.0001	<0.0001	0.515	0.045	0.205	0.857
	20–30	<0.0001	<0.0001	0.014	0.547	0.001	0.566	0.606
	30–40	<0.0001	0.025	<0.0001	0.137	0.012	0.001	0.414
NH_4^+–N	0–10	<0.0001	<0.0001	<0.0001	0.087	0.04	<0.0001	0.213
	10–20	<0.0001	0.004	<0.0001	0.292	0.046	0.002	0.809
	20–30	<0.0001	<0.0001	<0.0001	0.143	0.005	0.018	0.502
	30–40	<0.0001	<0.0001	<0.0001	0.051	0.036	<0.0001	0.516
NO_3^-–N	0–10	<0.0001	<0.0001	<0.0001	<0.0001	0.182	0.012	0.222
	10–20	<0.0001	<0.0001	<0.0001	<0.0001	0.082	<0.0001	0.580
	20–30	<0.0001	<0.0001	0.036	0.004	0.075	<0.0001	0.360
	30–40	<0.0001	<0.0001	<0.0001	0.037	0.746	0.117	0.924
TDP	0–10	<0.0001	0.015	<0.0001	0.211	0.01	0.418	0.734
	10–20	<0.0001	<0.0001	<0.0001	0.056	0.015	0.083	0.519
	20–30	<0.0001	<0.0001	<0.0001	0.248	0.047	0.01	0.807
	30–40	<0.0001	<0.0001	0.031	0.218	0.008	0.006	0.438

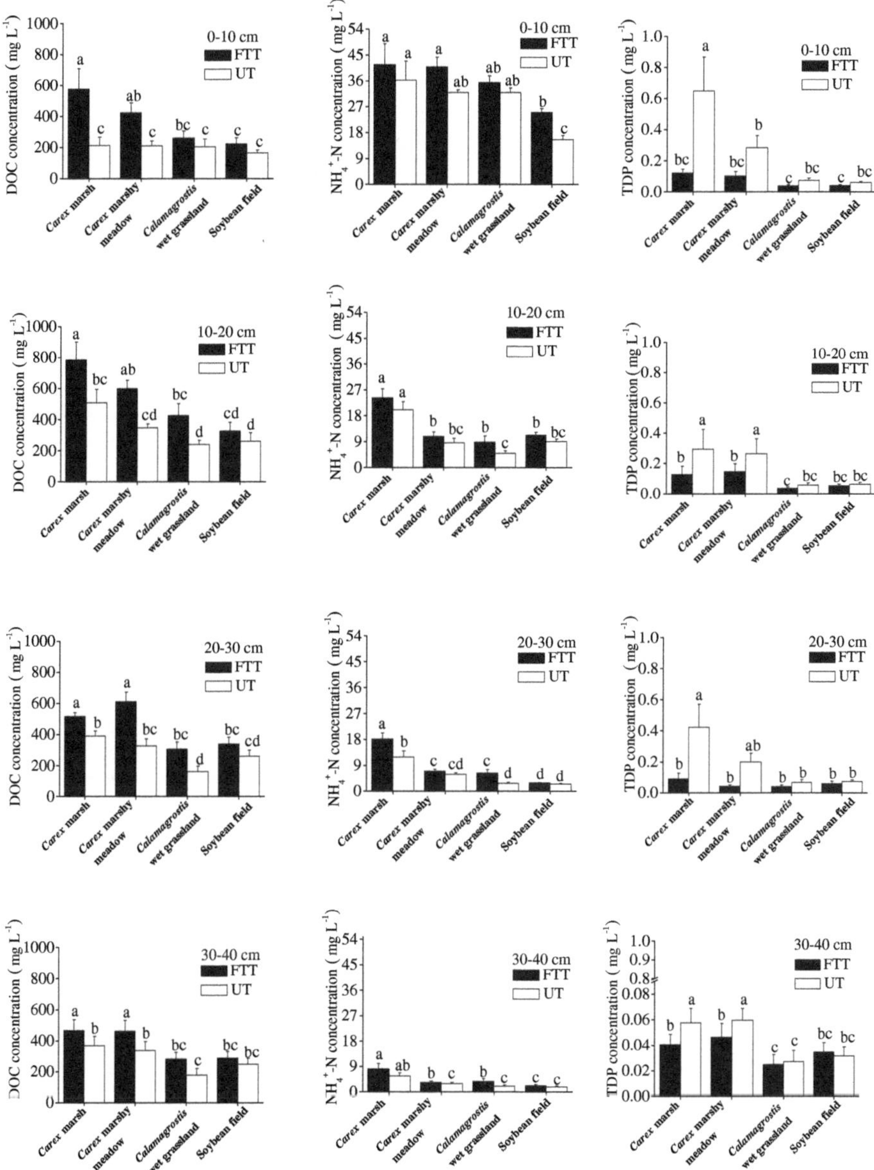

Fig. 8.1 Statistical differences of DOC, NH$_4^+$–N and TDP concentrations in soil solutions (Freeze–thaw treatment × Site interaction, mean ± standard error, $n = 21$). FTT and UT are soil columns with and without freeze–thaw treatment, respectively. *Carex* marsh, *Carex* marshy meadow, *Calamagrostis* wet grassland and soybean field are site types in the study (as follows). The site is identified on the x-axis, and the DOC, NH$_4^+$–N and TDP concentration in the soil solution is identified on the y-axis. Means with the same small letters are not significantly different at $p = 0.05$. (Reprinted from Yu et al. (2011), Copyright (2011), with permission from Elsevier)

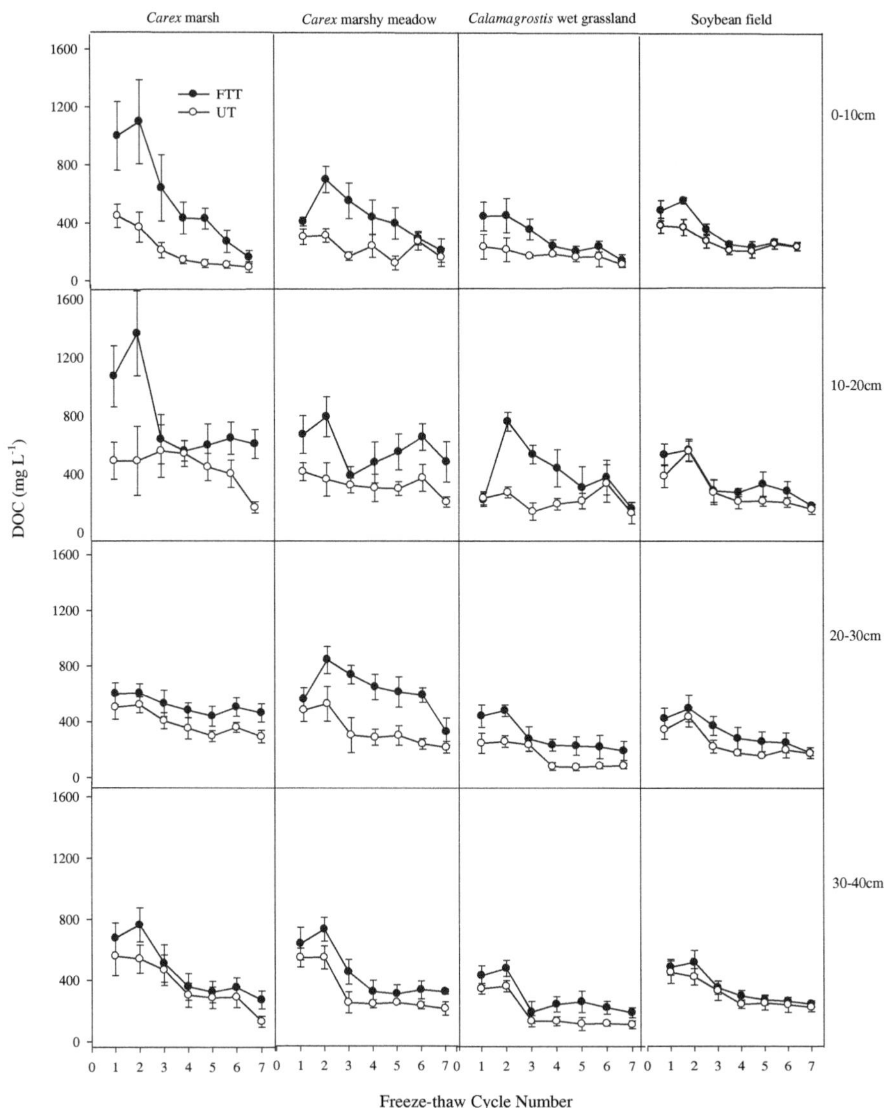

Fig. 8.2 DOC concentrations in soil solutions. FTT and UT are soil columns with and without freeze–thaw treatment, respectively. The length of time for each freeze–thaw cycle is 8 d. The number of freeze–thaw cycle is identified on the x-axis, and the DOC concentration in the soil solution is identified on the y-axis. Error bars represent the standard error of the mean of three parallel samples. (Reprinted from Yu et al. (2011), Copyright (2011), with permission from Elsevier)

and 20–30 cm) were greater than that in the 30–40 cm soil layer of the wetland soil columns (Fig. 8.3). The differences of average DOC increments between soil layers of the soybean field soil columns were not significant (Fig. 8.3). Compared to the DOC increments in the wetland soil columns, the DOC increments in the soybean field soil columns seemed small (ranging from 9 to 150 mg L^{-1}).

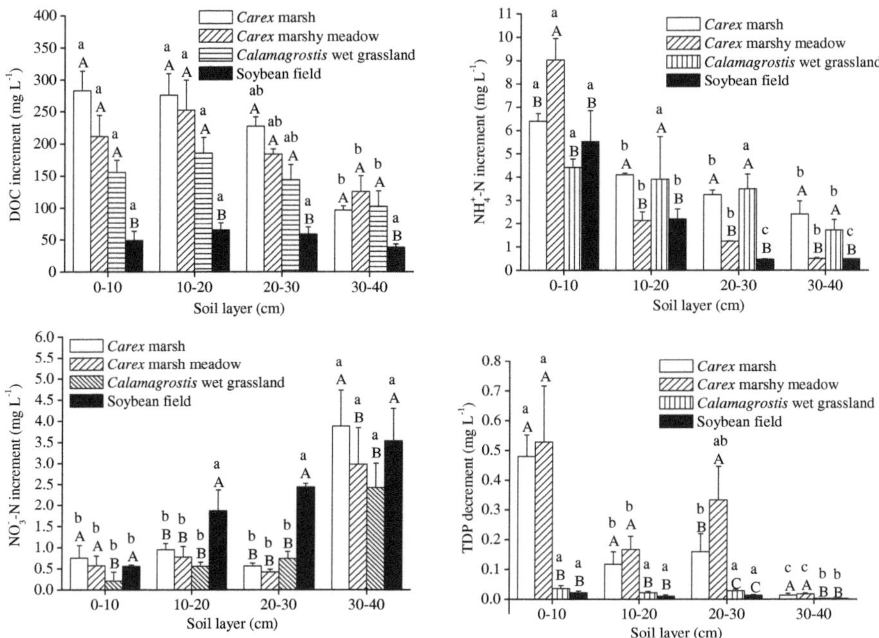

Fig. 8.3 Average increment of DOC, NH$_4^+$–N and NO$_3^-$–N concentrations and average decrement of TDP concentrations of seven cycles caused by freeze–thaw treatment (mean $\pm$ standard error, $n = 3$). The soil layer is identified on the x-axis, and the average increment of DOC NH$_4^+$–N and NO$_3^-$–N concentration and the average decrement of TDP concentrations of seven cycles caused by freeze–thaw treatment is identified on the y-axis. Means with the same capital letters within the same soil layer are not significantly different at $p = 0.05$. Means with the same small letters within the same sampling site are not significantly different at $p = 0.05$. Small standard errors are not visible. (Reprinted from Yu et al. (2011), Copyright (2011), with permission from Elsevier)

8.2   NH$_4^+$–N in Soil Solutions

The freeze–thaw treatment, cycle, soil layer and sampling site had significant effects on NH$_4^+$–N concentrations in soil solutions (Table 8.1, 8.2). The NH$_4^+$–N concentrations in soil solutions were not significantly affected by freeze–thaw treatment $\times$ cycle interaction (Table 8.1, 8.2). For *Carex* marsh, *Carex* marshy meadow and *Calamagrostis* wet grassland, the NH$_4^+$–N concentrations were not significantly affected by freeze–thaw treatment $\times$ soil layer interaction, indicating that soil layer had a negative effect on the change of NH$_4^+$–N concentrations caused by freeze–thaw treatment. For the soybean field, the NH$_4^+$–N concentrations were significantly affected by freeze–thaw treatment $\times$ soil layer interaction (Table 8.1). Meanwhile, freeze–thaw treatment $\times$ site interaction significantly affected the NH$_4^+$–N concentrations (Table 8.2, Fig. 8.1). In 107 of 112 paired data (freeze–thaw treatment and control), NH$_4^+$–N concentrations in soil solutions increased after freeze–thaw treatment (Fig. 8.4).

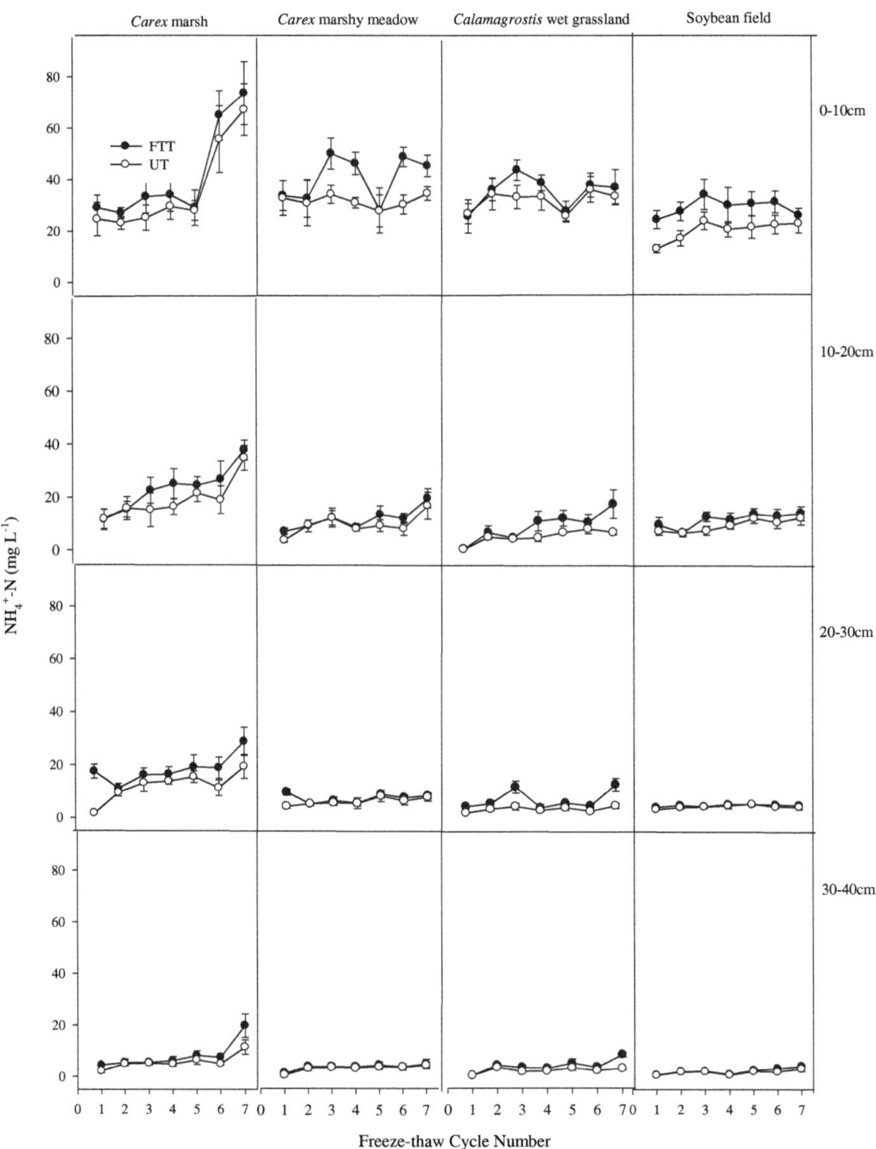

Fig. 8.4 NH_4^+–N concentrations in soil solutions. The number of freeze–thaw cycle is identified on the x-axis, and the NH_4^+–N concentration in the soil solution is identified on the y-axis. Error bars represent the standard error of the mean of three parallel samples. (Reprinted from Yu et al. (2011), Copyright (2011), with permission from Elsevier)

The NH_4^+–N increments caused by freeze–thaw treatment varying with cycles showed no consistent trend among various soil types or layers (Fig. 8.4). Therefore, we calculated the average NH_4^+–N increments of seven cycles, and compared the differences in various sampling sites and soil layers (Fig. 8.3). The average NH$_4{}^+$–N increments caused by freeze–thaw treatment decreased with the increase of soil depth, with the order of 0–10 > 10–20 > 20–30 > 30–40 cm (Fig. 8.3).

8.3 NO$_3{}^-$–N in Soil Solutions

The freeze–thaw treatment, cycle, soil layer and sampling site had significant effects on NO$_3{}^-$–N concentrations in soil solutions (Table 8.1, 8.2). The NO$_3{}^-$–N concentrations in soil solutions were not significantly affected by freeze–thaw treatment × site interaction and freeze–thaw treatment × soil layer interaction, but were significantly affected by freeze–thaw treatment × cycle interaction (Table 8.1, 8.2, Fig. 8.5). In 109 of 112 paired data (freeze–thaw treatment and control), NO$_3{}^-$–N concentrations in soil solutions increased after freeze–thaw treatment (Fig. 8.6).

The NO$_3{}^-$–N increments caused by freeze–thaw treatment varying with cycles showed no consistent trend among various soil types or layers, although there were much more NO$_3{}^-$–N increments after the early two or three cycles than after the last four cycles (Fig. 8.6). According to the average increments of seven cycles, three wetland soils had the same order of 0–10 cm $\approx$ 10–20 cm $\approx$ 20–30 cm < 30–40 cm, while the order of the soybean field soils was 0–10 cm < 10–20 cm $\approx$ 20–30 cm $\approx$ 30–40 cm (Fig. 8.3).

8.4 TDP in Soil Solutions

The freeze–thaw treatment, cycle, soil layer and sampling site had significant effects on TDP concentrations in soil solutions (Table 8.1, 8.2). The TDP concentrations in soil solutions were not significantly affected by freeze–thaw treatment × cycle interaction and freeze–thaw treatment × soil layer interaction (Table 8.1, 8.2), but were significantly affected by freeze–thaw treatment × site interaction (Table 8.2, Fig. 8.1). In 106 of 112 paired data (freeze–thaw treatment and control), we observed the decreased TDP concentrations in soil solutions after freeze–thaw treatment (Fig. 8.7).

The TDP decrements caused by freeze–thaw treatment varying with cycles showed no consistent trend among various soil types or layers (Fig. 8.7). Therefore, we calculated the average decrements of seven cycles for each soil layer and sampling site (Fig. 8.3). The average TDP decrements of *Carex* marsh and *Carex* marshy meadow were greater than that of *Calamagrostis* wet grassland and soybean field (Fig. 8.3).

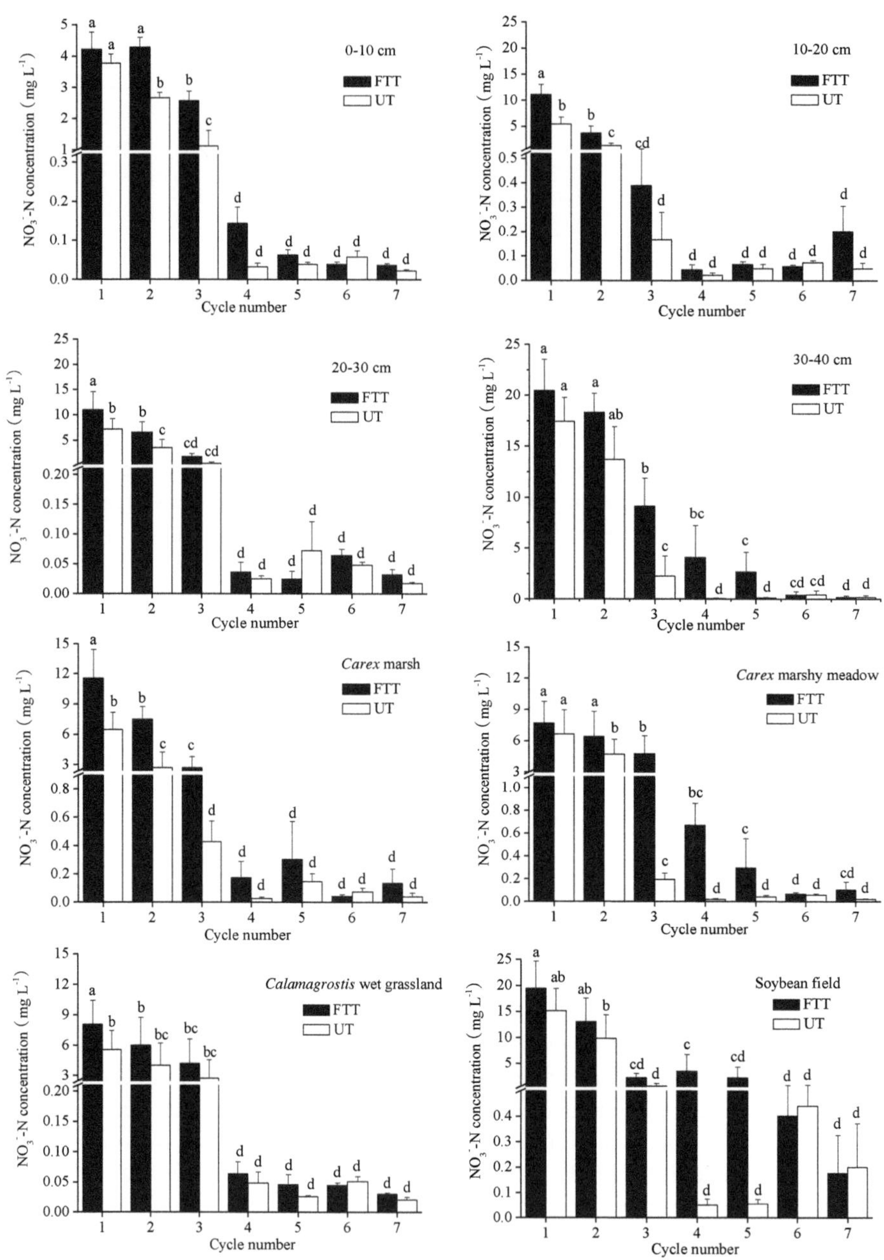

Fig. 8.5 Statistical differences of NO_3^-–N concentrations in soil solutions (Freeze–thaw treatment × cycle interaction, mean ± standard error, n = 12). The number of freeze–thaw cycle is identified on the x-axis, the NO_3^-–N concentration in the soil solution is identified on the y-axis. The upper graphs show the mean of all sites sorted by depth, and the lower ones show the mean of all depths within each site. Means with the same small letters are not significantly different at $p = 0.05$. (Reprinted from Yu et al. (2011), Copyright (2011), with permission from Elsevier)

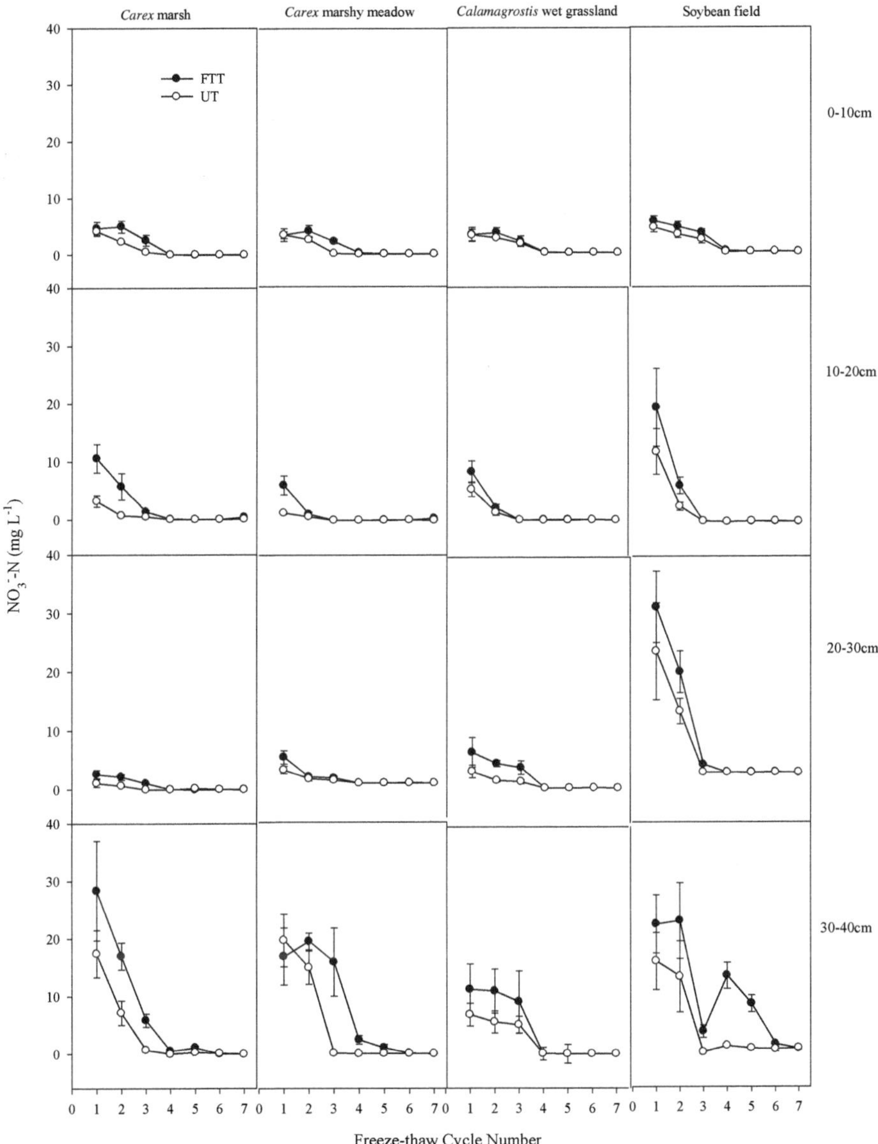

Fig. 8.6 NO$_3^-$–N concentrations in soil solutions. The number of freeze–thaw cycle is identified on the x-axis, and the NO$_3^-$–N concentration in the soil solution is identified on the y-axis. Error bars represent the standard error of the mean of three parallel samples. (Reprinted from Yu et al. (2011), Copyright (2011), with permission from Elsevier)

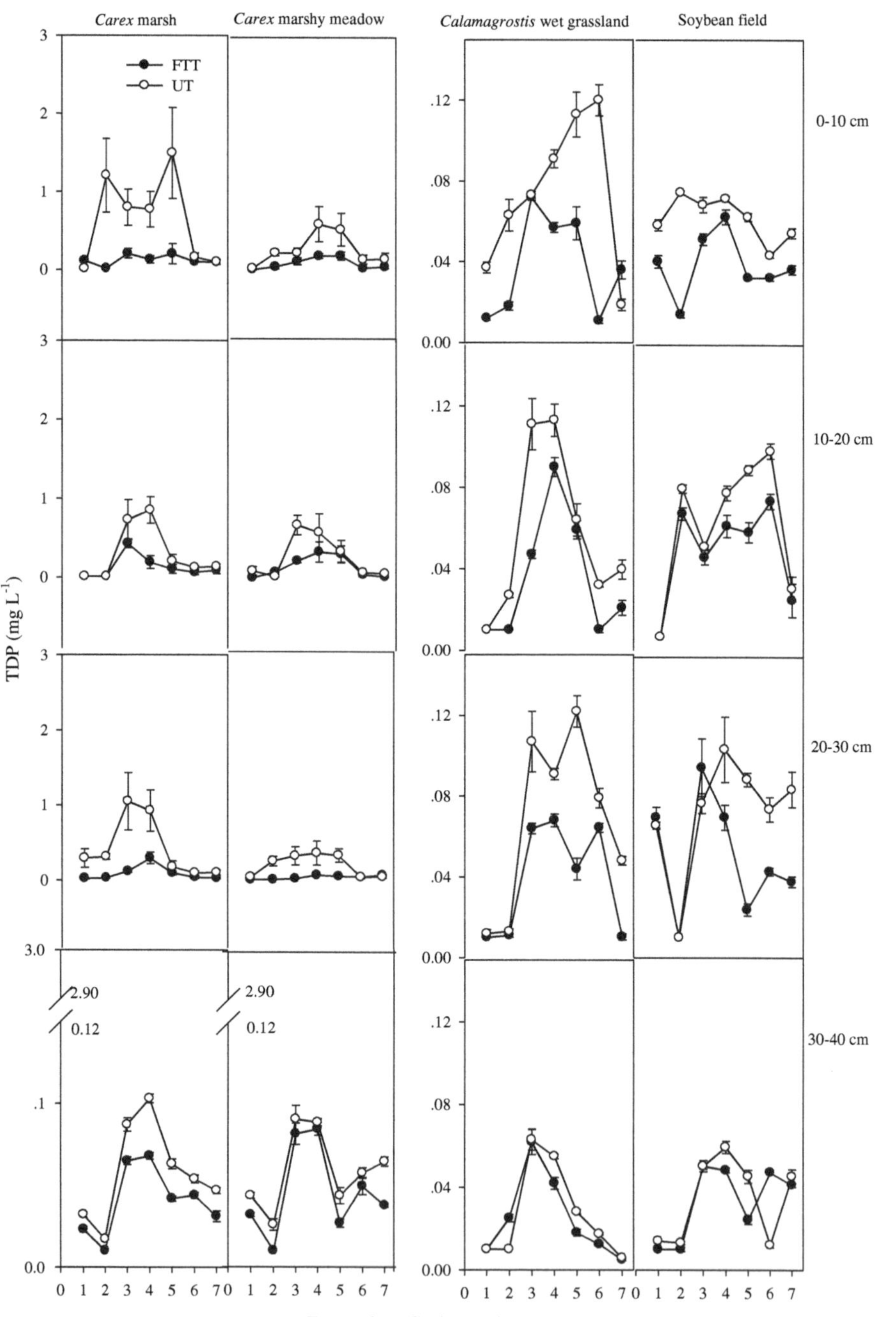

Fig. 8.7 TDP concentrations in soil solutions. The number of freeze–thaw cycle is identified on the x-axis, and the TDP concentration in the soil solution is identified on the y-axis. Error bars represent the standard error of the mean of three parallel samples. (Reprinted from Yu et al. (2011), Copyright (2011), with permission from Elsevier)

8.5 Freeze–Thaw Effects

The DOC concentrations in soil solutions were greater in treatment columns than in control columns (Fig. 8.2), indicating that freeze–thaw cycles could increase net DOC production of soils. It has been reported that DOC was released by killing of microbial cells, physical disruption of soil aggregates and root turnover during the freezing period (Ivarson and Sowden 1970; DeLuca et al. 1992; van Bochove et al. 2000; Tierney et al. 2001). These could be the explanations for the increase of DOC concentrations after freeze–thaw cycles. Skogland et al. (1988) and DeLuca et al. (1992) observed a single freeze–thaw cycle can even kill up to 50 % of the variable microbial populations, which subsequently resulted in the release of cell contents, including DOC. This explained the phenomenon in our experiment that an increase of DOC concentration occurred in the early two freeze–thaw cycles. The increase of DOC concentrations caused by freeze–thaw cycles might play an important role in the burst of N_2O emissions observed after thawing (Teepe et al. 2001; Sehy et al. 2004), and might also be relevant for the CO_2 emissions and the sequestration of soil organic carbon (Hentschel et al. 2008; Kalbitz and Kaiser 2008). In addition, in order to reduce soil disturbance, considerable dead or dying roots were introduced in the soil columns in our experiment, which might increase somewhat carbon and nitrogen in soil solutions.

The NH_4^+–N concentrations in soil solutions were greater in treatment columns than in control columns (Fig. 8.4), indicating that freeze–thaw cycles could increase net NH_4^+–N production of soils. Similar findings have been reported in some laboratory studies with forest soils and tundra soils (Grogan et al. 2004; Vestgarden and Austnes 2008). These could be attributed to microbial cell lysis, physical fragmentation of soil aggregates (Edwards 1991) and subsequent increase of microbial nitrogen mineralization after freeze–thaw cycles (Herrmann and Witter 2002). Moreover, increased NH_4^+–N concentrations in soil solutions could provide more NH_4^+–N for plant growth in the spring.

Similar with NH_4^+–N, the NO_3^-–N concentrations in soil solutions were greater in treatment columns than in control columns (Fig. 8.6). Nitrogen mineralization, nitrification and denitrification are three main biochemistry processes related to NO_3^-–N concentrations in soil solutions. Increased nitrogen mineralization, nitrification and denitrification after freeze–thaw cycles have been observed elsewhere (Christensen and Tiedje 1990; Christensen and Christensen 1991; Deluca et al. 1992; Herrmann and Witter 2002). Some researchers found that nitrogen mineralization decreased after freeze–thaw cycles (Larsen et al. 2002; Hentschel et al. 2008). Nitrogen mineralization and nitrification could lead to the production of NO_3^-–N, while denitrification could lead to the consumption of NO_3^-–N. As a result, the increase or decrease of NO_3^-–N concentrations after freeze–thaw cycles was determined by the changed production and consumption of NO_3^-–N after freeze–thaw cycles. In this study, we observed the increase of NO_3^-–N concentrations after freeze–thaw cycles, suggesting that the increased production of NO_3^-–N was greater than the increased consumption of NO_3^-–N caused by freeze–thaw cycles. Alternatively, the decrease of consumption could be bigger than the decrease of production. Therefore, there was the increase of net NO_3^-–N production in soils after freeze–thaw cycles. Flushes of NO_3^-–N in soils were also observed by some researchers in the field experiments during the early spring (Roth and Fox 1990; McCracken et al. 1994; Ryan et al. 2000). There is little plant uptake of NO_3^-–N in the early spring, and the increased NO_3^-–N concentrations in soil solutions present tremendous challenges from the water quality perspective.

The TDP concentrations in freeze–thaw treatment columns were smaller than that in control columns (Fig. 8.7), which indicated that freeze–thaw cycles could reduce the net release of TDP from soils. It was clear that the most prominent effect was an increase of TDP concentrations in the 2nd/3rd cycle (compared to 1st cycle, the timing varies, around 50 % of the columns showed an increase in the 2nd cycle already, the other half in the 3rd) (Fig. 8.7). The columns were assembled from discrete samples layer by layer, i.e. the natural condition was disturbed, and soil solution as present in situ was lost in the process. Then they were flooded with deionized water. Assuming constant net release of phosphorus, it would take some time before the concentration in the soil solution reached a new steady state, where release and uptake were balanced again. Although the columns have been flooded for 14 d before the beginning of the experiment, the release and uptake of phosphorus might not be balanced completely. In treatment columns, this recovery back to in situ concentration might be hindered by the freezing periods. Also Wang et al. (2007) suggested that desorption capacity of phosphorus was decreased by freeze–thaw cycles. These could explain the smaller TDP concentrations in freeze–thaw treatment columns.

8.6 Spatial Heterogeneity of the Freeze–Thaw Effects

The increments of DOC concentrations in soil solutions caused by freeze–thaw cycles were greater in the wetland soil columns (*Carex* marsh, *Carex* marshy meadow and *Calamagrostis* wet grassland) than in the soybean field soil columns (Fig. 8.3). These results in the 0–10 and 10–20 cm soil layers could be explained by the decreased microbial biomass carbon (MBC) in the 0–10 and 10–20 cm soil layers after the wetland reclamation (Zhang et al. 2006), because MBC is important to estimate the amount of DOC set free by cell lysis. Moreover, for the 20–30 and 30–40 cm soil layers, the DOC increments were also greater in the wetland soil columns than in the soybean field soil columns. In our experiment, many dead or dying roots were present in wetland soils, especially in the upper three soil layers of wetlands, while little were present in the 20–30 and 30–40 cm soil layers of the soybean field. Tierney et al. (2001) reported that some DOC was released from the deaths of roots and their subsequent decomposition after freeze–thaw cycles. These might explain the greater DOC increments in the 20–30 and 30–40 cm soil layers of the wetland soil columns than that of the soybean field soil columns. In addition, it has been suggested that the MBC was smaller in the deeper wetland soils (Barbhuiya et al. 2004), which might contribute to the greater DOC increments in the upper three soil layers than in the

30–40 cm soil layer of the wetland soil columns (Fig. 8.3). Furthermore, we observed that the DOC increments were positively related to the soil DOC contents and soil organic carbon contents ($R = 0.815$, $p < 0.001$; $R = 0.852$, $p < 0.001$, respectively).

The increments of NH_4^+–N concentrations in soil solutions caused by freeze–thaw cycles decreased with the increase of soil depth (Fig. 8.3), which might be attributed to the decreased microbial biomass nitrogen (Barbhuiya et al. 2004) and the decreased nitrogen mineralization with the increase of wetland soil depth (Hadas et al. 1989). Additionally, we found that the increments were positively related to the organic carbon content of soils ($R = 0.639$, $p = 0.008$), indicating that the freeze–thaw effects were linked to soil organic carbon content.

For the wetland soil columns (*Carex* marsh, *Carex* marshy meadow and *Calamagrostis* wet grassland), the increments of NO_3^-–N concentrations caused by freeze–thaw cycles in the upper three soil layers were smaller than that in the 30–40 cm soil layer (Fig. 8.3). For the soybean field soil columns, the increments in the 0–10 cm soil layer were smaller than that in the deeper soil layers (Fig. 8.3). An explanation might be that NO_3^-–N leached from the upper soils to the deeper soils, and hence the NO_3^-–N increments caused by freeze–thaw cycles consequently accumulated in the deeper soils. NO_3^-–N is negatively charged, and difficult to be adsorbed by soil colloids, which makes NO_3^-–N leaching easy. Although there was no vertical water flow in the soil columns, we speculated there was NO_3^-–N leaching because of gravity under the flooding condition. It was reported that NO_3^-–N leaching was easier in soils with less clay content and greater soil porosity (Gaines and Gaines 1994; Fan et al. 2010). Soil clay contents in the wetland soils were less than that in the soybean field soils (Table 8.3). Besides, soil

Table 8.3 Soil physical and chemical properties for natural wetland and farmed soils in the Sanjiang Plain, northeast China (mean values from three samples)

Column	Layer (cm)	DOC (mg kg^{-1})	NH_4^+–N (mg kg^{-1})	NO_3^-–N (mg kg^{-1})	Total-P (mg kg^{-1})	Organic carbon (%)	Clay content (%)
Carex	0–10	3879	1312	1.93	1644	23.45	15.46
marsh	10–20	2762	1146	1.64	1512	20.19	22.74
	20–30	2139	1104	1.06	1408	16.52	32.20
	30–40	1128	1036	2.76	1326	11.69	39.14
Carex	0–10	1725	823	0.64	1425	19.59	26.33
marshy	10–20	1109	665	0.56	1300	16.27	31.33
meadow	20–30	1024	542	0.60	1237	12.27	45.03
	30–40	888	210	0.39	1015	10.66	49.62
Calamagros-	0–10	546	362	1.90	1234	11.32	58.55
tis wet	10–20	353	268	1.16	1021	7.26	67.14
grassland	20–30	495	236	1.81	982	5.78	63.31
	30–40	551	133	1.51	840	1.15	63.79
Soybean	0–10	197	496	40.53	1195	10.19	60.00
field	10–20	194	427	38.11	990	3.18	69.17
	20–30	178	138	13.72	799	1.25	68.57
	30–40	118	141	7.57	686	0.57	70.90

porosities would be greater in the wetland soils than in the soybean field soils, for more roots were present in the wetland soils. Therefore, $NO_3^- - N$ leaching in the wetland soil columns might be easier than that in the soybean field soil columns, Consequently, $NO_3^- - N$ leaching could penetrate through the upper three soil layers of the wetland soil columns, while could only penetrate through the first soil layer of the soybean field soil columns, which might contribute to the smaller $NO_3^- - N$ increments in the upper three soil layers than in the 30–40 cm soil layer of the wetlands oil columns, and the smaller $NO_3^- - N$ increments in the 0–10 cm soil layer than in other soils layers of the soybean field soil columns.

Another explanation for the smaller $NO_3^- - N$ increments in the upper three soil layers of the wetland soil columns could be that the greater increased denitrification consumption of $NO_3^- - N$ occurred in the upper three soil layers than in the 30–40 cm soil layer. As mentioned in the discussion on the increased $NO_3^- - N$ caused by freeze–thaw cycles, the increased $NO_3^- - N$ concentrations after freeze–thaw cycles was determined by the increased production minus the increased denitrification consumption of $NO_3^- - N$ after freeze–thaw cycles. It was reported that the freeze–thaw-induced release of decomposable organic carbon was the major driving force for denitrification (Christensen and Christensen 1991; Morkved et al. 2006). We hypothesized that there were more increased consumption of $NO_3^- - N$ by denitrification after freeze–thaw cycles in the upper three soil layers of the wetland soil columns because of more freeze–thaw-induced release of decomposable organic carbon. In fact, the DOC increments caused by freeze–thaw cycles were greater in the upper three soil layers than in the 30–40 cm soil layer of the wetland soil columns (Fig. 8.3), which supported our hypothesis.

For the control columns, the TDP concentrations in soil solutions were much greater in *Carex* marsh and *Carex* marshy meadow than in *Calamagrostis* wet grassland and the soybean field (Fig. 8.3). The soil clay contents of *Calamagrostis* wet grassland and the soybean field were much greater than that of *Carex* marsh and *Carex* marshy meadow (Table 8.3), and soils with higher clay content had greater sorption capacity of phosphorus (Brennan et al. 1994), which might contribute to the less release of TDP in *Calamagrostis* wet grassland and the soybean field soil columns than in *Carex* marsh and *Carex* marshy meadow soil columns. The little release of TDP in soil solutions led to the limited TDP decrements after freeze–thaw cycles. As a result, the TDP decrements in *Calamagrostis* wet grassland and the soybean field soil columns were much smaller than that in *Carex* marsh and *Carex* marshy meadow soil columns.

8.7 Chapter Summary

The study demonstrated that freeze–thaw cycles can increase DOC, $NH_4^+ - N$ and $NO_3^- - N$ concentrations, and decrease TDP concentrations in soil solutions. Spatial heterogeneity of freeze–thaw effects related to DOC, $NH_4^+ - N$, $NO_3^- - N$ and TDP concentrations in soil solutions were apparent, further influenced by sampling site and soil layer. The freeze–thaw has a major influence on the nutrient release and retention of nutrients in wetland soils. The experimental approach used in this research can be generally applied in the study of soil solutions from semi-disturbed or undisturbed soil systems. Our study might have important implications related to water quality in the cold-winter zone.

References

Barbhuiya AR, Arunachalam A, Pandey HN, Arunachalam K, Khan ML, Nath PC. Dynamics of soil microbial biomass C, N and P in disturbed and undisturbed stands of a tropical wet-evergreen forest. Eur J Soil Biol. 2004;40:113–21.

Brennan RF, Bolland MDA, Jeffery RC, Allen DG. Phosphorus adsorption by a range of Western Australian soils related to soil properties. Commun Soil Sci Plant Anal. 1994;25:2785–95.

Buckeridge KM, Grogan P. Deepened snow alters soil microbial nutrient limitations in arctic birch hummock tundra. Appl Soil Ecol. 2008;39:210–22.

Christensen S, Christensen BT. Organic–matter available for denitrification in different soil fractions–effect of freeze/thaw cycles and straw disposal. J Soil Sci. 1991;42:637–47.

Christensen S, Tiedje JM. Brief and vigorous N_2O production by soil at spring fall. J Soil Sci. 1990;41:1–4.

DeLuca TH, Keeney DR, McCarty GW. Effect of freeze–thaw events on mineralization of soil nitrogen. Biol Fertil Soils. 1992;14:116–20.

Edwards LM. The effect of alternate freezing and thawing on aggregate stability and aggregate size distribution of some Prince Edward Island soil. J Soil Sci. 1991;42:193–204.

Fan J, Hao MD, Malhi SS. Accumulation of nitrate-N in the soil profile and its implications for the environment under dryland agriculture in northern China: a review. Can J Soil Sci. 2010;90:429–40.

Gaines TP, Gaines ST. Soil texture effect on nitrate leaching in soil percolates. Commun Soil Sci Plant Anal. 1994;25:2561–70.

Goldberg S, Muhr J, Borken W, Gebauer G. Fluxes of climate-relevant trace gases between a Norway spruce forest soil and the atmosphere during repeated freeze-thaw cycles in mesocosms. J Plant Nutr Soil Sci. 2008;171:729–39.

Grogan P, Michelsen A, Ambus P, Jonasson S. Freeze–thaw regime effects on carbon and nitrogen dynamics in sub–arctic heath tundra mesocosms. Soil Biol Biochem. 2004;36:641–54.

Hadas A, Feigin A, Feigenbaum S, Portnoy R. Nitrogen mineralization in the field at various depths. Eur J Soil Sci. 1989;40:131–7.

Henry HAL. Soil freeze–thaw cycle experiments: trends, methodological weaknesses and suggested improvements. Soil Biol Biochem. 2007;39:977–86.

Hentschel K, Borken W, Matzner E. Repeated freeze–thaw events affect leaching losses of nitrogen and dissolved organic matter in a forest soil. J Plant Nutr Soil Sci. 2008;171:699–706.

Herrmann A, Witter E. Sources of C and N contributing to the flush in mineralization upon freeze–thaw cycles in soils. Soil Biol Biochem. 2002;34:1495–505.

Ivarson KC, Sowden FJ. Effect of frost action and storage of soil at freezing temperatures on the free amino acids, free sugars and respiratory activity of soil. Can J Soil Sci. 1970;50:191–8.

Kalbitz K, Kaiser K. Contribution of dissolved organic matter to carbon storage in forest soils. J Plant Nutr Soil Sci. 2008;171:52–60.

Lipson DA, Monson RK. Plant–microbe competition for soil amino acids in the alpine tundra: effects of freeze–thaw and dry–rewet events. Oecologia. 1998;113:406–14.

Larsen KS, Jonasson S, Michelsen A. Repeated freezee−thaw cycles and their effects on biological processes in two arctic ecosystem types. Appl Soil Ecol. 2002;21:187-95.

Ludwig B, Teepe R, de Gerenyu VL, Flessa H. CO_2 and N_2O emissions from gleyic soils in the Russian tundra and a German forest during freeze-thaw periods–a microcosm study. Soil Biol Biochem. 2006;38:3516–9.

Matzner E, Borken W. Do freeze-thaw events enhance C and N losses from soils of different ecosystems. Eur J Soil Sci. 2008;59:274–84.

McCracken DV, Smith MS, Grove JH, et al. Nitrate leaching as influenced by cover cropping and nitrogen source. Soil Sci Soc Am J. 1994;58:1476–83.

Morkved PT, Dorsch P, Henriksen TM, Bakken LR. N_2O emissions and product ratios of nitrification and denitrification as affected by freezing and thawing. Soil Biol Biochem. 2006;38:3411–20.

Muller C, Martin M, Stevens RJ, et al. Processes leading to N_2O emissions in grassland soil during soil freezing and thawing. Soil Biol Biochem. 2002;34:1325–31.

Prieme A, Christensen S. Natural perturbations, drying–rewetting and freeze–thawing cycles, and the emission of nitrous oxide, carbon dioxide and methane from farmed organic soils. Soil Biol Biochem. 2001;33:2083–91.

Roth GW, Fox RH. Soil nitrate accumulations following nitrogen–fertilized corn in Pennsylvania. J Environ Qual. 1990;19:243–8.

Ryan MC, Kachanoski RG, Gillham RW. Overwinter soil nitrogen dynamics in seasonally frozen soils. Can J Soil Sci. 2000;80:541–50.

Schimel JP, Clein JS. Microbial response to freeze–thaw cycles in tundra and taiga soils. Soil Biol Biochem. 1996;28:1061–6.

Sehy U, Dyckmans J, Ruser R, Munch JC. Adding dissolved organic carbon to simulate freeze–thaw related N_2O emissions from soil. J Plant Nutr Soil Sci. 2004;167:471–8.

Skogland T, Lomeland S, Goksoyr J. Respiratory burst after freezing and thawing of soil: experiments with soil bacteria. Soil Biol Biochem. 1988;20:851–6.

Teepe R, Brumme R, Beese F. Nitrous oxide emissions from soil during freezing and thawing periods. Soil Biol Biochem. 2001;33:1269–75.

Tierney G, Fahey TJ, Groffman PM, Hardy JP, Fitzhugh RD, Driscoll CT. Soil freezing alters fine root dynamics in a northern hardwood forest. Biogeochemistry. 2001;56:175–90.

van Bochove E, Prevost D, Pelletier F. Effects of freeze–thaw and soil structure on nitrous oxide produced in a clay soil. Soil Sci Soc Am J. 2000;64:1638–43.

Vestgarden LS, Austnes K. Effects of freeze–thaw on C and N release from soils below different vegetation in a montane system: a laboratory experiment. Glob Change Biol. 2008;15:876–87.

Wang G, Liu J, Zhao H, Wang J, Yu J. Phosphorus sorption by freeze-thaw treated wetland soils derived from a winter-cold zone. Geoderma. 2007;138:153–61.

Yu XF, Zou YC, Jiang M, et al. Response of soil constituents to freeze-thaw cycles in wetland soil solution. Soil Biol Biochem. 2011;43(6):1308–20.

Zhang JB, Song CC, Yang WY. Land use effects on the distribution of labile organic carbon fractions through soil profiles. Soil Sci Soc Am J. 2006;70:660–7.

Zhao H, Zhang X, Xu S, Zhao X, Xie Z, Wang Q. Effect of freezing on soil nitrogen mineralization under different plant communities in a semi-arid area during a non-growing season. Appl Soil Ecol. 2010;45:187–92.

Zou YC, Lu XG, Jiang M. Dynamics of dissolved iron under pedohydrological regime caused by pulsed rainfall events in wetland soils. Geoderma. 2009;150:46–53.

Chapter 9
Conclusions and Prospect

9.1 Conclusions

Freeze–thaw can promote the production and fixation of DOC, NH_4^+–N and NO_3^-–N in wetland soils, and promote the depletion and release of wetland soil organic carbon and organic nitrogen. The DOC, NH_4^+–N and NO_3^-–N concentrations in the water overlying wetland and downstream water bodies will increase in the freeze–thaw period. As a result of global warming, fixation of DOC and NH_4^+–N and the release of organic carbon in wetland soils may be weakened. Wetland reclamation also promotes the release of DOC and NH_4^+–N from wetland soils.

9.1.1 The Freeze–Thaw Rules of Wetland Soils

The Sanjiang Plain belongs to the seasonal freeze–thaw area. The freezing period of this area is long. The temperature decreases significantly in autumn. The soils begin to be frozen in October, and begin to be thawed. The freezing period can last six months. There are seasonal variation and diurnal variation in the freeze–thaw process. The complex and significant freeze–thaw process must have effects on the biogeochemistry process of wetlands. In this study, by doing the in situ observation for two years, we analyzed the temperature variations of soils, overlying waters and snow. We also simulated the responses of freeze–thaw regime to the global climate warming. The results showed that the simulated global warming by covered plastic films speeded up the thawing of overlying snow and ice of wetland soils, which affected the thawing of the covered soils indirectly. Under the same snow precipitation, the increased surface water depth as a result of acceleration of snow and ice melts, affected the water temperature and freeze–thaw process of the covered soils. When the wetlands were reclaimed, the soil temperature was increased or decreased acutely.

X. Yu, *Material Cycling of Wetland Soils Driven by Freeze–Thaw Effects*,   107
Springer Theses, DOI: 10.1007/978-3-642-34465-7_9,
© Springer-Verlag Berlin Heidelberg 2013

9.1.2 The Dynamics of Dissolved Carbon and Nitrogen in Wetland Soils, Overlying Ice and Overlying Water During the Freeze–Thaw Period

Freeze–thaw can affect soil physical and chemical properties, microbial activities and nutrition cycling. And the dynamics of dissolved carbon and nitrogen contents in wetland soils, overlying ice and overlying water during the freeze–thaw period are the integrated reflects of these processes. In this study, we did the semi-simulation experiments in field in order to use the natural freeze–thaw condition in field. During the spring freeze–thaw period, DOC (dissolved organic carbon), NH_4^+–N and NO_3^-–N accumulations in wetland soils, the overlying ice and overlying water varied regularly. DOC, NH_4^+–N and NO_3^-–N accumulations in wetland soils were increased. DOC, NH_4^+–N and NO_3^-–N concentrations in the overlying water in the spring freeze–thaw period were all greater than background concentrations in the last autumn.

9.1.3 The Freeze–Thaw Effects on the Sorption/Desorption of DOC

The sorption of DOC is not only the major control for organic matter (OM) and OM-assisted transport; it also contributes to the stabilization and accumulation of organic matter in soils. Freeze–thaw cycles have disruptive effects on soils structure, which decrease bulk density and penetration resistance and increase the specific surface area. Freeze–thaw cycles increased the sorption capacity of DOC and reduced desorption potential of DOC in wetland soils and reclaimed wetland soils. The freeze–thaw effects on desorption potential of DOC could be improved by increasing FTCs. The wetland reclamation reduced the sorption capacity of DOC and increased the desorption potential of DOC.

9.1.4 The Freeze–Thaw Effects on the Sorption/Desorption of NH_4^+

The NH_4^+ is the primary form of mineralized nitrogen in most flooded wetland soils. Because of the anaerobic nature of wetland soils, NH_4^+ can be prevented from further oxidation and is stable. The sorption/desorption of NH_4^+ is an important mechanism of N retention and release in wetland soils. The sorption/desorption of NH_4^+ is related to N availability for plant uptake and N leaching processes in soils. Freeze–thaw significantly increased the sorption capacity of NH_4^+ and reduced the desorption potential of NH_4^+ in wetland soils. There was significant difference in the NH_4^+ adsorbed amount between soils with and without freeze–thaw treatment. The adsorbed amount of NH_4^+ increased with the increasing freeze–thaw cycles. The depressional wetland soils had greater sorption capacity and weaker desorption potential of NH_4^+ than riparian wetland soils resulted from the higher clay content and CEC. Because of the altered soil physical and chemical properties and hydroperiod, sorption capacity of NH_4^+ was weaker in farmland soils than in wetland soils while desorption potential was stronger in farmland soils.

9.1.5 The Freeze–Thaw Effects on the Organic Carbon and Organic Nitrogen Mineralization of Wetland Soils

Soil organic carbon and organic nitrogen mineralization is the most important process in soil carbon and nitrogen cycles. The freeze–thaw effects on soil organic carbon and organic nitrogen mineralization depends on the balance of the decreased mineralization because of the decreased microbial biomass and the increased mineralization results from the lysis of dead microorganisms. In this study, we did some simulation experiments under freeze–thaw condition. The results showed that freeze–thaw increased the mineralization amount of wetland soil organic carbon (SOC) obviously, enhanced the degree of SOC mineralization, and made it easier. Dryland soil had greater mineralization amount than wetland soils. Of the wetland soils, that of *Carex lasiocarpa* marsh was greater than that of *Calamagrostis angustifolia* marshy meadow. Freeze–thaw increased the mineralization amount of wetland soil organic nitrogen (SON) significantly, too, with greater effect on *Carex lasiocarpa* marsh soil than *Calamagrostis angustifolia* marshy meadow and dryland soils. *Carex lasiocarpa* marsh soil had greater mineralization amount, which was positively correlated with the DOC concentration in soils.

9.1.6 The Dynamics of Accumulation and Release of Dissolved Carbon and Nitrogen in Intact Wetland Soils

Chemical changes in the soil solutions of soils over time often reflect physicochemical processes such as dissolved matter transfer and transport. In previous research of freeze–thaw effects on C and N concentrations in soil solutions, the soil solutions were extracted by K_2SO_4 solution, but the process can destroy the chemical composition and the equilibrium of the soil solutions. In fact, an in situ soil solution sampler is required to do the experiments without disrupting the soil structure. In this study, we incubated intact wetland soil columns and simulated freeze–thaw effects on DOC, NH_4^+–N and NO_3^-–N concentrations in wetland and farmland soil solutions, and investigated the spatial heterogeneity of the freeze–thaw effects. The results showed that freeze–thaw can significantly increase DOC, NH_4^+–N and NO_3^-–N concentrations in soil solutions. And the freeze–thaw treatment can also affect the concentration varying trends. Besides, there were spatial heterogeneities of the freeze–thaw effects on DOC, NH_4^+–N and NO_3^-–N concentrations in soil solutions.

9.2 Shortness and Outlook of the Research

In a three-year preliminary study on the Sanjiang Plain, freeze–thaw effects on the accumulation and release of wetland soil carbon and nitrogen were explored. While interesting preliminary results were obtained, there are still some limitations which require further research.

9.2.1 The Introduction of Biogeochemical Models

This thesis has focused on the indoor and outdoor simulation of the main processes involved in carbon and nitrogen accumulation and release in wetland soils under freeze–thaw conditions, supplemented by observations of the freeze–thaw process in wetland and farmland soils. This has revealed the mechanism of the freeze–thaw effect on the accumulation and release of soil carbon and nitrogen. However, we cannot predict the carbon and nitrogen cycle of the entire wetland ecosystem during the freeze–thaw period. The next step should be to combine with biogeochemical models of the wetland ecosystem, such as the DNDC model, to study carbon and nitrogen accumulation and release processes in wetland soils in the freeze–thaw period, thus more deeply improving our understanding of the impact of the freeze–thaw process on wetland carbon and nitrogen cycles.

9.2.2 The Mineral Analysis and Microbial Identification

The simulation results of freeze–thaw effects on wetland soil carbon and nitrogen adsorption and desorption, the mineralization of organic carbon and organic nitrogen in wetland soil describe phenomena, but the related mechanism study lacks clear mineral and microbial evidence. In the next step, analysis of the impact of freeze–thaw on soil minerals should be urgently addressed, as well as the isolation and identification of the freeze–thaw-affected microorganisms, to develop convincing mechanisms and create new hypotheses.

9.2.3 The Coupled Study of Field Simulation and Lab Experiment

Climate change is a popular topic of current study, and the freeze–thaw process in seasonal freeze–thaw regions is an environmental characteristic which is closely related to climate change. It is still unknown how changes in freeze–thaw temperatures and frequency because of climate change will affect wetland soil carbon and nitrogen accumulation and release. In the next step, field observations and laboratory simulation experiments should be combined to simulate the effects of different freeze–thaw situations on wetland soil carbon and nitrogen accumulation and release under climate warming conditions to reveal the response of wetland soil carbon and nitrogen cycles to climate warming in freeze–thaw regions during the seasonal freeze–thaw period.